PHYSICS OF DENSE MATTER

PHYSICS OF DENSE MATTER

Y. C. Leung

Professor of Physics
Southeastern Massachusetts University

Science Press
Beijing, China

World Scientific
Singapore

Published by

Science Press
Beijing China

and

World Scientific Publishing Co. Pte. Ltd.
5 Toh Tuck Link, Singapore 596224
USA office: 27 Warren Street, Suite 401-402, Hackensack, NJ 07601
UK office: 57 Shelton Street, Covent Garden, London WC2H 9HE

British Library Cataloguing-in-Publication Data
A catalogue record for this book is available from the British Library.

Responsible Editor: Zhang Banggu

PHYSICS OF DENSE MATTER

ISBN-13 978-9971-978-10-5
ISBN-10 9971-978-10-5

Science Press Book No. 3445-42

To My Parents

PREFACE

This book grew from lecture notes prepared for a ten-week summer short-course on dense matter physics that I gave to a group of students and physicists at the Beijing Normal University (China) in 1979. My aim was to introduce the field of dense matter physics which saw rapid development in the past ten to fifteen years to an audience who was not yet active in the field. I wanted to make the subject matter as simple and yet as complete as I could by concentrating on only a few well-developed topics, and I tried to spell out all details as clearly as possible. My hope was to provide enough background information on these topics that the participants of the short-course might be prepared to comprehend the current research articles. In reality, I was far from reaching my goal, and so in 1981 when I was approached by the Science Press to publish the lecture notes, I decided to rewrite them. The final result comprises the present book, which contains fewer topics but more detailed and complete description of each.

This book is intended to prepare the readers for theoretical studies in dense matter physics. It may also be used pedagogically as an introduction to the many-body theory with emphasis on illustrative examples. Many mathematical operations needed for this field are carried out in details, usually not in the most elegant way but in a practitioner's way, so that readers who are handicapped in background training can still follow. Hopefully this book will enable more newcomers to enter into the study of dense matter.

I wish to thank my colleagues at the Beijing Normal University for comments and helpful suggestions on the contents of this book. Among them I wish to mention the following persons: Liang Shao-rong, Li Zong-wei, Gao Shang-hui, Ge Yun-zao and Shi Tian-yi, whom I had the pleasure to work with during the preparation of this book. Finally, I wish to thank Chen Chung-kuang of the Graduate School of the Academy of Sciences for arranging with the Science Press for me to publish this book.

Y.C. Leung
N. Dartmouth, Massachusetts
May 4, 1983

CONTENTS

INTRODUCTION

In this book an attempt is made to provide a theoretical background for the study of the structure of very dense matter. The densities under investigation are those between 10^4 to 10^{16} g/cm^3 found in inert stellar objects like the white dwarfs and neutron stars. Even though dense matter physics is also related closely to contemporary investigations in heavy ion collisions and the Early Universe, we shall make little reference to these issues. Instead the emphasis is placed on the stellar objectes with the final aim in deriving the appropriate equation of state for matter forming them. Thus, not all aspects of dense matter physics are dealt with but just the structure of dense matter.

All macroscopic solids on earth have average densities low compared with the range of densities mentioned above. Their structures are of electrostatic in origin. We shall say very little about them. We begin our study at densities above 10^4 g/cm^3, for which electrostatic effects play minor roles. At these densities quantum effects and nuclear interactions occupy the central roles. These are the topics under review here.

The average density of our Sun is very low only of the order of 1.5 g/cm^3. This is because energy coming from thermal nuclear reaction prevents the collapse of matter to high density. Only the helium core at the center of the Sun is reaching high density, but the core is still small due to the youthfulness of our Sun.

As a star ages its helium core grows in size. The support of the helium core is due to degenerate electrons, which in a dense state can resist considerable pressure. The relationship between density and pressure for a substance is called its equation of state. Once the equation of state is known the mass and radius of the core can be determined. The maximum stable size of such a core is known to be given by the Chandrasekhar limit, which is about 1.2 solar mass. Chapter 1 deals with topics which are related to the inert core below the Chandrasekhar limit. Matter in the core is composed of nuclei whose atomic numbers are below that of iron. We refer to such substance as subferrous matter. In Chapter 1 general many-body techniques like the Hartree and Hartree-Fock methods are introduced. Relevant concepts in dealing with degenerate fermions, particle interactions and thermal effects are reviewed for the familiar case of the electron gas. It serves to establish the notations and general method of approach.

What then is the fate of a star whose helium core exceeds the Chandrasekhar limit? A sequence of events would have followed leading eventually to a possible supernova process. There are considerable astrophysical interests in knowing the detailed nature of the supernova process for it will lead to an understanding of the chemical composition of our Solar System. Although some of the basic mechanisms of the supernova process are known, many more fine details are needed to complete the picture

Chapter 2 makes an attempt to understand the physical properties of matter during the process of a gravitational collapse. The results shown in Chapter 2 are far from complete and undoubtedly a great deal of progress in understanding matter at such densities will be made in the near future. Chapter 2 serves to introduce the methods presently employed to tackle such problems. In presenting these methods we have not try to cover all possible approachs but concentrate on a few which are phenomenological and show promise for future development. We try to indicate their strength and weakness whenever we can. Chapter 2 deals with matter at densities above those of the white dwarf stars but below that found in atomic nuclei. It is comparable to a gas of nuclei whose atomic numbers are above that of iron and is referred to as subnuclear matter. The transformation in matter's composition with density requires careful analysis. A definitive knowledge of the process will be crucial to an understanding of the atomic abundance of the Solar System.

Matter at subnuclear densities is quite compressible. It will not halt a gravitational collapse once it is initiated. Stellar objects involved in a collapse must necessarily pass through a stage where their core densities are as high as matter densities found inside atomic nuclei. Densities in atomic nuclei are quite similar and are usually specified by a single density called the nuclear density. Gravitational collapse will compress matter density well beyond the nuclear density. The halting of a gravitational collapse is due to the structure of matter at transnuclear densities. Consequently, knowledge of matter at such densities is crucial to the understanding of the supernova process. In Chapter 3 we introduce techniques in nuclear physics appropriate to the study of matter at approximately the nuclear density. The purpose is to provide the general background needed to understand many of the research articles published in this field. We try to work out all the mathematical details involved in the theory, so that newcomers to this field will not be hindered by obscure derivations. We leave out however philosophical comments on such theory since they can be found in severalgood review articles quoted at the end of each section. The historical sketch on the development of the theory is also left out. In Section 14, we introduce relativistic models which allows the extension of nuclear physics techniques to deal with matter at densities above nuclear density. Chapter 3 covers a density range terminating at about 10^{16} g/cm^3.

Matter at densities above 10^{16} g/cm^3 is called ultradense matter. Physical processes not commonly found in nuclear matter will appear in ultradense matter. The discussion of these processes composes the subject matter of Chapter 4. They are listed under the headings of baryonic matter, pion condensation and quark matter. Recognition of these processes is largely theoretical. We try to present these issues in a unified way without having to introduce drastically new devices. Since the tone of this book is highly phenomenological, many of the sophisticated diagrammatic techniques have been neglected, and therefore it is impossible to give a full treatment of some of the topics.

There are three appendices to this book. Appendix A compiles the physical constants and astronomical data which will be useful for the subject matter. Appendix B presents the mathematical functions employed in the text. Appendix C is a compendium of the equation of state for matter at various densities. It represents our present day knowledge of such matter. Exercises are provided for each section and are given at the end of the book.

CHAPTER 1

Regime I: $500 < \rho < 10^8$ g/cm^3. Subferrous Matter.

1. Variational Methods

The study of the physical properties of matter by means of theoretical methods comes under a formulation generally referred to as the many-body problem, in which matter is assumed to consist of a number of basic constituents called particles obeying known physical laws. All physical properties of matter are to be deduced from the dynamics of these constituents. The quantum mechanical formulation of the many-body problem is based on a generalization of the rather successful Schrodinger equation formulation of the one- and two-body problems. Thus, we write for a N-body system the following equation:

$$H \Psi = E \Psi \tag{1.1}$$

where Ψ is a function of the 3N spatial coordinates $\vec{x}_i$, and the Hamiltonian H is given by:

$$H = \sum_{i=1}^{N} -\frac{\hbar^2}{2m} \nabla_i^2 + \sum_{i=1}^{N} V(\vec{x}_i) + \sum_{\substack{i,j=1 \\ i<j}}^{N} v(\vec{x}_i - \vec{x}_j) , \tag{1.2}$$

which consists of the kinetic energy terms and the interaction terms. The particles are assumed to interact with the background through $V(\vec{x}_i)$ and pairwise through $v(\vec{x}_i-\vec{x}_j)$. In a quantum mechanical formulation of particle dynamics, each particle is provided with its own spatial coordinates, and hence there are 3N coordinates. Interactions between two particles are prescribed by means of $v(\vec{x}_i-\vec{x}_j)$, which couples their coordinates. Without $v(\vec{x}_i-\vec{x}_j)$, (1.1) will be separable into N sets of independent coordinates and all particles are then free from each other. We have in mind here a system of identical particles and there will be the same potential functions $V(\vec{x}_i)$ and $v(\vec{x}_i-\vec{x}_j)$ for all particles. For the general case additional subscripts for V and v will be needed. These subscripts are suppressed here for clarity.

It is clearly a very ambitious task to solve for Ψ from (1.1). Furthermore, when the system represents a macroscopic system, the determination of Ψ is not only impossible but not necessary. For such a system, the process of a physical measurement samples some average characteristics of the system, and this is far from a determination of the many-body wave function Ψ. However, the concept of a wave function is still useful as a vehicle in establishing relations among physical parameters. Henceforth, we shall attack the problem in a manner consistent with an approximate solution of the many-body problem (1.1).

One common strategy in finding the ground state energy, or the lowest eigenvalue of (1.1), is by means of the variational method. The variational principle tells us that the expectation value of the Hamiltonian with an arbitrary function must be at least as great as the lowest expectation value of the Hamiltonian. One therefore specifies a class of trial wave functions which are presumed variable functionally from one to the other. The variational method then provides the necessary equations for the selection of the one trial wave function which gives the lowest expectation value of the Hamiltonian within the class. Such a trial wave function may be a useful approximation to the true ground state wave function if the class is properly chosen, and its expectation value would be a close upper bound of the ground state energy of the system. Different classes of trial wave functions may also be appraised by comparing their lowest expectation values obtained. We shall make use of three classes of trial wave functions. They shall be designated as (1) Hartree, (2) Hartree-Fock, and (3) Jastrow trial wave functions.

After the approximate ground state wave function is found for the variational method. The excited state wave functions may be established successively. The first excited state wave function must be orthogonal to the ground state wave function and at the same time yielding the next lowest expectation value of the Hamiltonian. The second excited state wave function would again be orthogonal to the first two with the next lowest eigenvalue, and so forth. We shall be dealing mainly with the ground state of a system.

A N-particle Hartree trial wave function consists of a product of N single-particle wave functions:

$$\Psi = \phi_1(\vec{x}_1)\cdot \phi_2(\vec{x}_2) \;.\;.\;.\;.\;.\;.\; \phi_N(\vec{x}_N) \;. \tag{1.3}$$

The single-particle wave functions ϕ_i are individually normalized:

$$\int d^3x \;\; \phi_i^*(\vec{x})\, \phi_i(\vec{x}) = 1, \tag{1.4}$$

so that,

$$\int d^Nv \;\; \Psi^* \Psi = 1 \,, \tag{1.5}$$

where

$$d^Nv = d^3x_1 \, d^3x_2 \;.\;.\;.\;.\;.\;.\;.\; d^3x_N \;. \tag{1.6}$$

Let us denote the expectation value of the Hamiltonian by:

$$< H > = \int d^Nv \;\; \Psi^* H \Psi \;. \tag{1.7}$$

A variation in the trial wave function Ψ: $\Psi \to \Psi + \delta\Psi$ gives rise to a variation in $<H>$: $<H> \to <H> + \delta<H>$. Since Ψ is complex, Ψ and Ψ^* may be varied independently. The variation should however preserve the normalization conditions (1.5). This can be accomplished by means of the method of Lagrange multiplier. The variational condition to be satisfied by the trial wave function having the lowest expectation value of the Hamiltonian is:

$$\delta \{ < H > + \lambda \int d^Nv \, \Psi^* \Psi \} = 0 \,, \tag{1.8}$$

where λ is the Lagrange multiplier.

Since the single-particle wave functions are normalized, $<H>$ is given by:

$$\begin{aligned} < H > = & \sum_i \int d^3x \, \phi_i^*(\vec{x}) \, \{-\frac{\hbar^2}{2m}\nabla^2 + V_i(\vec{x})\} \, \phi_i(\vec{x}) \; + \\ & + \sum_{i<j} \int d^3x \, d^3y \; \phi_i^*(\vec{x}) \; \phi_j^*(\vec{y}) \; v(\vec{x}-\vec{y}) \; \phi_i(\vec{x}) \; \phi_j(\vec{y}) \;. \end{aligned} \tag{1.9}$$

Let us consider variations of Ψ^*: $\Psi^* \to \Psi^* + \delta\Psi^*$ by varying just one of the single-particle wave functions at a time, such as:

$$\delta\Psi^* = \delta\phi_1^*(\vec{x}_1)\cdot \phi_2^*(\vec{x}_2) \;.\;.\;.\;.\;.\;.\; \phi_N^*(\vec{x}_N) \;. \tag{1.10}$$

The variational condition,

$$\int d^Nv \, \{ \; \delta\Psi^* \; (H + \lambda) \;\; \Psi \; \} = 0 \tag{1.11}$$

simplifies to:

$$\int d^3x\, \delta\phi_1^*(\vec{x}) \{ -\frac{\hbar^2}{2m}\nabla^2 + V(\vec{x}) + \sum_{j=2}^{N} \int d^3y\, \phi_j^*(\vec{y})\, v(\vec{x}-\vec{y})\, \phi_j(\vec{y}) +$$

$$+ (\lambda + \sum_{j=2}^{N} b_j + \sum_{1<\ell<j} c_{\ell j})\}\, \phi_1(\vec{x}) = 0 , \quad (1.12)$$

where,

$$b_j = \int d^3x\, \phi_j^*(\vec{x}) \{-\frac{\hbar^2}{2m}\nabla^2 + V(\vec{x})\}\, \phi_j(\vec{x}) , \quad (1.13)$$

and,

$$c_{\ell j} = \int d^3x\, d^3y\, \phi_\ell^*(\vec{x})\, \phi_j^*(\vec{y})\, v(\vec{x}-\vec{y})\, \phi_\ell(\vec{x})\, \phi_j(\vec{y}) . \quad (1.14)$$

since $\delta\phi_1^*$ is an arbitrary complex function, the vanishing of the integral can only be satisfied by the vanishing of the integrand. Thus, we obtain a Hartree equation for the single-particle wave function labelled by subscript 1. This may be generalized for arbitrary particle label i. The equations obtained are the Hartree equations, one for each particle wave function:

$$\{-\frac{\hbar^2}{2m}\nabla^2 + V(\vec{x}) + \sum_{\substack{j=1 \\ (j\neq i)}}^{N} \int d^3y\, \phi_j^*(\vec{y}) v(\vec{x}-\vec{y}) \phi_j(\vec{y})\}\, \phi_i(\vec{x}) = \varepsilon_i^H\, \phi_i(\vec{x}) , \quad (1.15)$$

where,

$$\varepsilon_i^H = - (\lambda + \sum_{\substack{j=1 \\ (j\neq i)}}^{N} b_j + \sum_{\substack{\ell,j=1 \\ (\ell<j) \\ (\ell,j\neq i)}}^{N} c_{\ell j})$$

are the Hartree single-particle energies. The approximate ground state energy of the system, denoted again by E, is given by the expectation value of the Hamiltonian:

$$E = < H > = \sum_{i=1}^{N} \int d^3x\, \phi_i^*(\vec{x}) \{-\frac{\hbar^2}{2m}\nabla^2 + V(\vec{x})\}\, \phi_i(\vec{x}) +$$

$$+ \sum_{i<j}^{N} \int d^3x\, d^3y\, \phi_i^*(\vec{x})\, \phi_j^*(\vec{y})\, v(\vec{x}-\vec{y})\, \phi_i(\vec{x})\, \phi_j(\vec{y})$$

$$= \sum_{i=1}^{N} \varepsilon_i^H - \sum_{i<j} \int d^3x\, d^3y\, \phi_i^*(\vec{x})\, \phi_j^*(\vec{y})\, v(\vec{x}-\vec{y})\, \phi_i(\vec{x})\, \phi_j(\vec{y}) . \quad (1.16)$$

Here we see that the total energy of the system is given by the sum of the Hartree single-particle energies minus the interaction energies, which have been doubly counted in the summation of the Hartree single-particle energies. The Hartree single-particle energy corresponds physically to the removal energy of a particle from the system and shall be referred to as the removal energy. Also, we note that each Hartree equation for wave function ϕ_i depends on the knowledge of all other wave functions ϕ_j. Hence, in solving for ϕ_i one demands self-consistency in the sense that the function ϕ_j used for the evaluation of ϕ_i should be the same as that obtained when ϕ_i is used as an input in the evaluation of ϕ_j. The self-consistency requirement may be accomplished by means of iterations of the trial wave functions. It turns out that in practice the caluculation is not as bad as it might have seemed, largely because the dependence of the interaction terms on the wave functions is through combinations $\Sigma\phi_j^* \phi_j$ which has the interpretation of particle distribution and may be estimated with fair accuracy from the start.

Hartree-Fock Equations

The Hartree trial wave functions are strictly speaking not appropriate for the treatment of identical particles, which obey either Bose statistics in which case the total wave function of these particles should be totally symmetric under the interchange of the particle coordinates, or Fermi statistics in which case the total wave function should be totally antisymmetric. We shall be dealing mainly with the case of Fermi statistics such as a system of electrons or nucleons. Trial wave functions for identical fermions have been given by Fock and they will be referred to as the Hartree-Fock trial wave functions. They consist of products of single-particle wave functions properly antisymmetrized. A Hartree-Fock trial wave function of N particles may be written in the form of a Slater determinant of single-particle wave functions ψ_i as follows:

$$\Psi = \frac{1}{\sqrt{N!}} \begin{vmatrix} \psi_1(\vec{x}_1) & \psi_1(\vec{x}_2) & \cdots & \psi_1(\vec{x}_N) \\ \psi_2(\vec{x}_1) & \psi_2(\vec{x}_2) & \cdots & \psi_2(\vec{x}_N) \\ \psi_3(\vec{x}_1) & \cdots & \cdots & \cdots \\ \cdots & \cdots & \cdots & \cdots \\ \psi_N(\vec{x}_1) & \cdots & \cdots & \psi_N(\vec{x}_N) \end{vmatrix} \qquad (1.17)$$

where the square array of single-particle wave functions ψ_i is to be evaluated in the manner of a determinant. Such linear combinations of single-particle wave functions would possess the antisymmetric property demanded for identical fermions.

Since each spin-half fermion possesses two spin components, called spin-up and spin-down, it is necessary to include a spin wave function describing the spin state of each single-particle wave function. The spin wave function will be written as $\chi_\sigma(i)$, and its conjugate as $\chi_\sigma^+(i)$, where σ denotes $\uparrow$ or $\downarrow$ for the spin-up or spin-down state, respectively. The spin wave functions are orthonormal:

$$\chi_\sigma^+(i) \quad \chi_{\sigma'}(i) = \delta(\sigma,\sigma') \tag{1.18}$$

where $\delta(\sigma,\sigma')$ denotes the Kronecker delta symbol. Hence, each single-particle wave function decomposes into:

$$\psi_i(\vec{x}) \quad = \quad \phi_i(\vec{x}) \; \chi_\sigma(i) \tag{1.19}$$

where $\phi_i(\vec{x})$, as in the Hartree equations, depend only on the spatial coordinates. The conjugate wave functions will be written as $\psi_i^+(\vec{x}) = \phi_i^*(\vec{x}) \; \chi_\sigma^+(i)$. For a Hamiltonian H describing a system of identical particles it is invariant under the exchange of the particle coordinates: $\vec{x}_i \leftrightarrow \vec{x}_j$, and consequently all potential functions $V(\vec{x}_i)$ or $v(\vec{x}_i-\vec{x}_j)$ have the same form.

A variation on the total wave function Ψ^+ may be conveniently obtained by writing $\delta\Psi^+$ in the form of (1.17) with ψ_1^+ replaced by $\delta\psi_1^+$. The variation of the matrix elements of the Hamiltonian may be simplified as follows:

$$\delta\Psi^+ \; H \; \Psi \quad = \{\delta\psi_1^+(\vec{x}_1) \; \psi_2^+(\vec{x}_2) \; . \; . \; . \; \psi_N^+(\vec{x}_N)\} \quad H \; \{\sqrt{N!} \; \Psi \} \tag{1.20}$$

where we have performed N! permutations of the coordinates to reduce the determinantal form of $\delta\Psi^+$ into a single term. The above operation takes advantage of the invariance of the Hamiltonian under such permutations. $\{\sqrt{N!} \; \Psi\}$ consists then just the determinantal part and without the factorial normalization factor.

Let us examine the expectation values of the Hamiltonian in two parts: the first part involves the one-particle operators and the second part the two-particle operators:

(a) For one-particle operators: $\{ f_i \equiv -\frac{\hbar^2}{2m}\nabla_i^2 + V(\vec{x}_i) \}$

$$\int d^N v\ \delta\Psi^+ f_i\ \Psi = \int d^3x_1\ \delta\psi_1^+(\vec{x}_1)\ f_1\ \psi_1(\vec{x}_1)\ , \quad (\text{for } i = 1),$$

$$= \int d^3x\ \ \delta\psi_1^+(\vec{x})\ \psi_1(\vec{x}) \int d^3x_i\ \ \psi_i^+(\vec{x}_i)\ f_i\ \psi_i(\vec{x}_i)$$

$$= b_i \int d^3x\ \delta\psi^+(\vec{x})\ \psi(\vec{x}), \quad (\text{for } i \neq 1). \tag{1.21}$$

(b) For two-particle operators: $\{ v_{ij} \equiv v(\vec{x}_i - \vec{x}_j)\}$

$$\int d^N v\ \delta\Psi^+\ v_{ij}\ \Psi = \int d^3x d^3y\ \delta\psi_1^+(\vec{x})\ \psi_j^+(\vec{y})\ v_{xy}\ \{\psi_1(\vec{x})\psi_j(\vec{y}) - \psi_j(\vec{x})\psi_1(\vec{y})\},$$

$$(\text{for } i = 1,\ j > 1),$$

$$= d_{ij}\ \int d^3x\ \delta\psi_1^+(\vec{x})\ \psi_1(\vec{x})\ , \quad (\text{for } i,j > 1), \tag{1.22}$$

where

$$d_{ij} = \int d^3x\ d^3y\ \psi_i^+(\vec{x})\ \psi_j^+(\vec{y})\ v_{xy}\ \{\ \psi_i(\vec{x})\psi_j(\vec{y}) - \psi_j(\vec{x})\psi_i(\vec{y})\}$$

In summary, the variational condition may be expressed as:

$$\int d^N v\ \{\ \delta\Psi^+\ (\ H + \lambda\ )\ \Psi\ \} =$$

$$= \int d^3x\ \delta\psi_1^+(\vec{x})\ \{ \Big(-\frac{\hbar^2}{2m}\nabla^2 + V(\vec{x}) + \sum_{j>1} \int d^3y\ \psi_j^+(\vec{y}) v_{xy}\psi_j(\vec{y}) \Big)\ \psi_1(\vec{x}) +$$

$$- \sum_{j>1} \int d^3y \psi_j^+(\vec{y}) v_{xy}\psi_1(\vec{y})\ \psi_j(\vec{x}) + (\ \lambda + \sum_{j>1} b_j + \sum_{\ell>j>1} d_{\ell j})\ \psi_1(\vec{x}) \} = 0. \tag{1.23}$$

Through the vanishing of the integrand, we obtain the Hartree-Fock variational equation for wave function labelled by subscript 1. Further, by writing the single-particle wave function ψ_i in the form (1.19) the spin wave functions may be removed:

$$\psi_i^+(\vec{y})\ \psi_j(\vec{y}) = \phi_i^*(\vec{y})\ \phi_j(\vec{y})\ \chi_{\sigma_i}^+(i)\ \ \chi_{\sigma_j}(j) = \phi_i^*(\vec{y})\ \phi_j(\vec{y})\ \delta(\sigma_i,\sigma_j). \tag{1.24}$$

We are then left with a set of Hartree-Fock equations for the determination of the spatial dependent part of the wave functions as follows:

$$\{-\frac{\hbar^2}{2m}\nabla^2 + V(\vec{x}) + \sum_{j=1}^{N}\int d^3y\ \phi_j^*(\vec{y})v(\vec{x}-\vec{y})\phi_j(\vec{y})\}\ \phi_i(\vec{x}) +$$

$$-\sum_{j=1}^{N}\delta(\sigma_i,\sigma_j)\int d^3y\ \phi_j^*(\vec{y})v(\vec{x}-\vec{y})\phi_i(\vec{y})\ \phi_j(\vec{x}) = \varepsilon_i^{HF}\ \phi_i(\vec{x})\ , \qquad (1.25)$$

where the Hartree-Fock single-particle energies are given by:

$$\varepsilon_i^{HF} = -\ (\ \lambda + \sum_{j\neq i} b_j + \sum_{\substack{\ell<j \\ \ell,j\neq i}} d_{\ell j}\ )\ . \qquad (1.26)$$

Individually, ε_i^{HF} corresponds to the removal energy of the ith fermion from the system. In (1.25) both summations over j are unrestricted, since the j=i terms are automatically eliminated due to antisymmetry in the i and j indices.

The extra terms possessed by the Hartree-Fock equations over the Hartree equations are called the exchange terms, which account for the antisymmetry of the total wave function. The total energy E of the system is given in terms of the Hartree-Fock single-particle energies by:

$$E = \sum_i \varepsilon_i^{HF} - \sum_{i<j}\int d^3x\ d^3y\ \phi_i^*(\vec{x})\phi_j^*(\vec{y})\ v(\vec{x}-\vec{y})\{\ \phi_i(\vec{x})\phi_j(\vec{y}) - \delta(\sigma_i,\sigma_j)\phi_j(\vec{x})\phi_i(\vec{y})\} \qquad (1.27)$$

The interaction energies would have to be substracted from the sum of the single-particle energies since they have been counted twice in the sum.

We shall first work with the Hartree and the Hartree-Fock equations and postpone the discussion of the Jastrow trial wave functions until Chapter 3.

Bibliography

Hartree, D.R. (1957). *The Calculation of Atomic Structures*, Wiley, New York.

Landau, L.D. and Lifshitz, E.M. (1965). *Quantum Mechanics: Non-Relativistic Theory*, Addison-Wesley, Reading, Mass.

March, N.H. (1975). *Self-Consistent Fields in Atoms*, Pergamon, New York.

Slater, J.C. (1960). *Quantum Theory of Atomic Structure*, Vol. 2, McGraw-Hill, New York.

Slater, J.C. (1974). *The Self-Consistent Fields for Molecules and Solids*, McGraw-Hill, New York.

2. Ideal Degenerate Electron Gas

The most elementary exercise in many-body physics is the case of a system of electrons for which all interactions may be neglected, and which are at a temperature sufficiently cold so that thermal energy may also be neglected. At the conclusion of the exercise we can then examine precisely under which conditions may interactions and thermal energy be so neglected.

We shall study situations where electrons are uniformly distributed over the entire volume and are so densely pack that exclusion effects of identical fermions come fully into play. For the system in its ground state all low lying energy states will be occupied by two electrons, one for each spin state. Both electrons will have the same energy and are said to be degenerate with each other. A system of non-interacting electrons which occupy all low lying energy states in degenerate pairs will be called an ideal degenerate electron gas. Such a situation represents a true ground state configuration of the system and is only realizable when thermal energy is absent. Thermal energy will excite some of the electrons to higher energy states leaving some of the lower lying energy states not fully degenerate. Such configurations will be called partial degeneracy.

Without interactions the Hartree-Fock (or Hartree) equations reduce to:

$$-\frac{\hbar^2}{2m}\nabla^2\phi_i(\vec{x}) = \varepsilon_i\ \phi_i(\vec{x})\ , \tag{2.1}$$

and the total energy of the system is simply $E = \sum_i \varepsilon_i$. In addition to these equations there are also boundary conditions on ϕ_i to establish the eigenvalues ε_i.

The wave functions ϕ_i are not confined by a potential and they represent states of the electrons in a uniform medium, which being uniform is independent of where it is examined. This means the wave functions ϕ_i should possess translational invariance, meaning that it is the eigenfunction of the translational operator. A translational operator $T(\vec{\ell})$ performs the following operation on a wave function $\phi(\vec{x})$:

$$T(\vec{\ell})\ \phi(\vec{x}) = \phi(\vec{x} + \vec{\ell})\ . \tag{2.2}$$

It can readily be verified that $\phi(\vec{x}) = C\exp(i\vec{k}.\vec{x})$ is an eigenfunction of $T(\vec{\ell})$ for:

$$T(\vec{\ell})\ \phi(\vec{x}) \quad = \quad C\ \exp\{i\vec{k}\cdot(\vec{x}+\vec{\ell})\} \quad = \quad \exp(i\vec{k}\cdot\vec{\ell})\ \phi(\vec{x}) \ ,$$

is an eigenfunctional relation with eigenvalue given by $\exp(i\vec{k}\cdot\vec{\ell})$.

The form of a single-particle wave function for a uniform system may be specified completely by this invariance property, and is given as follows:

$$\phi_i(\vec{x}) \quad = \quad (\Omega)^{-\frac{1}{2}} \exp(i\ \vec{p}\cdot\vec{x}/\hbar) \ , \tag{2.3}$$

where $\hbar$ is as before the Planck's constant divided by 2π. The vector quantity $\vec{p}$ is to be interpreted as the momentum of the particle in that state according to the rules of quantum mechanics. The factor $(\Omega)^{-\frac{1}{2}}$ is added for the normalization of the wave function in a volume Ω. The wave function when restricted to a finite volume should obey periodic boundary conditions, or in other words, within a cubic volume of size $\Omega = \ell^3$, say, the wave function satisfies:

$$\phi_i(x,y,z) \quad = \quad \phi_i(x+n_x\ell,\ y+n_y\ell,\ z+n_z\ell) \ , \tag{2.4}$$

where we write $\vec{x} = (x,y,z)$ and integers $n_x,\ n_y,\ n_z = 0, \pm 1, \pm 2, \pm 3, \ldots$ Correspondingly, the allowed momentum components for the states are:

$$p_x = n_x h/\ell \ , \qquad p_y = n_y h/\ell \ , \qquad p_z = n_z h/\ell \ . \tag{2.5}$$

where $\vec{p} = (p_x,p_y,p_z)$. Each ϕ_i is to be associated with a definite triplet of integers (n_x,n_y,n_z) which denotes a specific quantum state for an electron. The energy of an electron in such a state is given by (2.1):

$$\varepsilon_i \quad = \quad \frac{p_i^2}{2m} \quad = \quad \frac{h^2}{2m\ell^2}\ (\ n_x^2 + n_y^2 + n_z^2\ ) \tag{2.6}$$

For a system of electrons in its ground state, the low energy single-particle states first start to fill up beginning with (n_x,n_y,n_z) = (1,0,0), (0,1,0), (0,0,1), (1,1,0), etc, accommodating two electrons per momentum state. Since different sets of (n_x,n_y,n_z) may give the same energy (2.6), the degeneracy of the system is actually much higher than two per energy state. The term degenerate electrons would also refer to degeneracy of this kind. The highest single-particle energy reached by the ground state configuration of the electron gas is called its Fermi energy, ε_F, and the corresponding momentum its Fermi momentum, $p_F = (2m\varepsilon_F)^{\frac{1}{2}}$.

In order to find the total number of quantum states whose energy is equal to or less than the Fermi energy, let us imagine each (n_x, n_y, n_z) corresponds to a point in a cubic lattice with unit lattice spacing. Then the total number of such states is equal to the total number of lattice points within a spherical volume of radius R in such a lattice, where R is given by: $R = p_F \ell/h$. Since the total number of lattice points in the volume is given by $(4\pi R^3/3)$, the total number of states is:

$$\text{(number of states)} = \left(\frac{\ell^3}{h^3}\right) \frac{4\pi}{3} p_F^3 ,$$

which is just the product of the spatial volume and the volume of the occupied momentum space divided by h^3. It is therefore convenient to imagine the situation simply as if the quantum states are being distributed uniformly over the momentum space, and for a ground state of the system, such states center around the origin of the momentum space.

The total number of electrons N in Ω is found by multiplying the total number of states computed from formula above by two to account for the two spin orientations possessed by each electron, and to remove the arbitrariness in choosing Ω, it is best to speak of an electron number density, n:

$$n = \frac{N}{\Omega} = \frac{2}{h^3} \int_0^{p_F} 4\pi\, p^2\, dp = \frac{2}{h^3} \left(\frac{4\pi}{3} p_F^3\right) \tag{2.7}$$

(2.7) may in turn be used to determine p_F once n is given.

With the kinetic energy of each state given by (2.6) and the states are distributed uniformly over the momentum space, the total ground state energy E of the system per unit volume, or its energy density, denoted by ε, is evaluated to be:

$$\varepsilon = \frac{E}{\Omega} = \frac{2}{h^3} \int_0^{p_F} 4\pi\, p^2\, dp \left(\frac{p^2}{2m}\right) = \frac{4\pi\, p_F^5}{5m\, h^3} \tag{2.8}$$

The pressure P of the electron gas can be evaluated from simple kinetic theory arguments which go as follows: the pressure exerted by a gas is the mean rate of transfer of momentum across an imaginary surface of unit area in the gas, and for an electron of momentum p_x and velocity v_x in a cubic volume Ω it imparts an average pressure of $p_x v_x / \Omega$ on the walls of volume which are perpendicular to the x-axis. The pressure from all the electrons is then given by:

$$P = (2/h^3) \int dp_x \, dp_y \, dp_z \, (p_x v_x) \; . \qquad (2.9)$$

From isotropy, $p_x v_x = (1/3)(p_x v_x + p_y v_y + p_z v_z) = (1/3)(\vec{p} \cdot \vec{v}) = pv/3$, and (2.9) reduces to

$$P = \frac{2}{3h^3} \int_0^{p_F} 4\pi \, p^2 dp \, (pv) \; . \qquad (2.10)$$

Since

$$v = \frac{\partial}{\partial p} \left(\frac{p^2}{2m} \right) = \frac{p}{m} \; , \qquad (2.11)$$

thus,

$$P = \frac{8\pi}{3mh^3} \int_0^{p_F} dp \, p^4 = \frac{8\pi \, p_F^5}{15mh^3} \; . \qquad (2.12)$$

The pressure P so calculated is indeed the pressure of the electron gas as can be verified by applying thermodynamics.

In thermodynamics the pressure P is derivable from the total energy E of the system as follows:

$$P = - \left. \frac{\partial E}{\partial \Omega} \right|_N \; . \qquad (2.13)$$

In this case, E is given by (2.8) and N by (2.7). From (2.13) we derive:

$$P = - \frac{8\pi}{10mh^3} \left\{ p_F^5 + 5 \, \Omega \, p_F^4 \frac{\partial p_F}{\partial \Omega} \right\} \; , \qquad (2.14)$$

where $\frac{\partial p_F}{\partial \Omega}$ is to be found from (2.7) for a constant N:

$$0 = \frac{4\pi}{3h^3} \left\{ p_F^3 + 3 \, \Omega \, p_F^2 \frac{\partial p_F}{\partial \Omega} \right\} \; ,$$

or,

$$\frac{\partial p_F}{\partial \Omega} = - \frac{p_F}{3\Omega} \; . \qquad (2.15)$$

Together, they give:

$$P = \frac{8\pi}{10mh^3} \left\{ -p_F^5 - 5 \, \Omega \, p_F^4 \left(- \frac{p_F}{3\Omega} \right) \right\} = \frac{8\pi \, p_F^5}{15mh^3} \; , \qquad (2.16)$$

which is the same as the result in (2.12).

Since pressure P is inversely proportional to the mass of the particle, it is clear that the partial pressure due to a proton gas at the same number density as the electron gas would be smaller by a factor of $(m_p/m_e) \simeq 2000$ as compared to that of the electron gas, and may therefore be neglected. That is the reason why our discussions revolve around the electron gas only. The electrons being 2000 times lighter than the protons also turn relativistic at relatively low energy. We next turn our attention to the treatment of relativistic electrons.

We note that even though the Hartree-Fock equations are based on a non-relativistic Schrodinger formulation of the quantum mechanical many-body problem, the numeration of quantum states of an ideal degenerate electron gas is actually based on the boundary conditions imposed on a translationally invariant wave function. The distribution of the quantum states in the momentum space for a relativistic electron gas is unchanged. The only modification needed would be in the expression of the relativistic kinetic energy in terms of the momentum. For a non-interacting system the Hamiltonian commutes with the momentum operator, and their eigenvalues obey classical kinematics:

$$e_K = (p^2c^2 + m^2c^4)^{1/2} - mc^2 . \tag{2.17}$$

Thus, the (kinetic) energy density ε_K of a system of relativistic ideal degenerate electrons is given by:

$$\varepsilon_K = \frac{2}{h^3}\int_0^{p_F} 4\pi p^2 dp\, e_K , \tag{2.18}$$

and the pressure:

$$P = \frac{2}{3h^3}\int_0^{p_F} 4\pi p^2 dp\, (p\frac{\partial e_K}{\partial p}) \tag{2.19}$$

with the Fermi momentum p_F given in terms of the number density n by (2.7). In summary, we have:

$$n = \frac{8\pi}{3}(\frac{t}{\lambda})^3 , \tag{2.20}$$

$$\varepsilon_K = \frac{\pi mc^2}{\lambda^3}\{t(2t^2+1)\sqrt{t^2+1} - \ln(t+\sqrt{t^2+1})\} , \tag{2.21}$$

$$P = \frac{\pi mc^2}{\lambda^3}\{\frac{1}{3}t(2t^2-3)\sqrt{t^2+1} + \ln(t+\sqrt{t^2+1})\} , \tag{2.22}$$

where $\lambda = h/mc$ is 2π multiple of the Compton wavelength of the particle and $t = p_F/mc$ is dimensionless.

Returning to the non-relativistic situation, we see that the average energy per electron can be found from (2.7) and (2.8) to be:

$$\frac{E}{N} = 0.6 \left(\frac{p_F^2}{2m}\right) \propto n^{2/3} , \qquad (2.23)$$

which increases rapidly with the electron number density n. We shall refer to this form of energy arising purely as a result of the exclusion principle and not due to interaction as degeneracy energy.

The average thermal energy per particle is given typically by the quantity k_BT, where k_B is the Boltzmann constant and T the equilibrium temperature. For an ordinary star, such as our Sun, at the stage of hydrogen burning, its interior temperature is of the order of 10^7K, or:

$$k_BT \simeq 1.5 \times 10^{-9} \text{ erg} \simeq 10^3 \text{ eV} . \qquad (2.24)$$

Degenerate electrons with average kinetic energy comparable to this thermal energy must reach number density of the order of $n \approx 3 \times 10^{26}$ $1/cm^3$. If we consider all the electrons to come from a fully ionized hydrogen gas, the mass density ρ of the proton-electron plasma is:

$$\rho = (m_p+m_e)\, n = 500 \text{ g/cm}^3 . \qquad (2.25)$$

Such density is high for most stars which are in the early stage of hydrogen burning, but for ageing stars which have developed a substantial helium core, the comparable density of $\rho = (m_n+m_p+m_e)n = 1000$ g/cm^3 can easily be exceeded within the core. Stellar objects like the white dwarfs have interior densities as high as 10^7 g/cm^3. In the study of the structure of the white dwarfs, the thermal energies become negligible and degeneracy energy plays the primary role.

The equation of state is a relation between the pressure P and the mass density ρ of the matter with definite composition and temperature. The equations of state which are of special interest in astrophysics in Regime I (mass density between 500 and 10^8 g/cm^3) are those for systems composing of plasmas of electrons and atomic nuclides ranging from helium to iron. A main sequence star will develop a helium core as hydrogen burning progresses. The stability of the helium core is crucial to the future development of the star. If the star is sufficiently small so that gravitational contraction of the core is insufficient to raise temperature high enough to ignite helium burning, the fully developed core will eventually turn into a white dwarf star. The stability of the core and the structure of the white dwarf star is determined by the equation of state of helium matter. On the other hand, if the core is sufficiently big and the collapse of the core ignites helium burning, then it is most likely that nuclear burning would be complete and the end point of the nuclear exothermic reactions is reached, in which case a core composing of ferrous nuclei begins to build up. The eventual collapse of the ferrous core is crucial to the understanding of the supernova phenomena. Hence the equation of state for ferrous matter is also of great interest. The equation of state pertinent to supernova process will be studied in the next chapter. In this section the equation of state for helium matter shall be worked out. So far, all interactions and thermal effects are neglected. These as we shall see in the next sections constitute only minor corrections, especially in the high density range of Regime I.

Since the pressure of the system in the present consideration is due entirely to degenerate electrons, the first thing to do is to establish the electron number density n of the system. Let the mass density of the matter be ρ, then the nucleon number density of the system is given by (ρ/m_p). It is customary to introduce the quantity Y_e which is the average number of electrons per nucleon in the system. Then, it is apparent that:

$$n = Y_e (\rho/m_p) \tag{2.26}$$

For a fully ionized helium plasma, $Y_e = 0.5$, and for a plasma of iron nuclei Y_e is given by the charge-to-mass number of the ferrous nuclide, $Y_e = 26/56 = 0.464$. Since Y_e for these two cases are very nearly the same, the equations of state for them should be also nearly identical. The equations of state for matter composing of electrons and nuclides between helium and iron will fall between these two, and therefore the equation of state for helium matter should be representative of this class. The astrophysical implication of this fact is that if a star is unstable against the equation of state of helium matter, it will also be unstable against all the others, and thus all possible exothermic reactions involving transmutations of nuclides below iron would ensue in the process. Consequently, a star which has gone beyond the white dwarf stage must necessarily enter the supernova stage without any possibility of intermediate existence.

The equation of state of helium matter at zero temperature is shown in Figure 2-1, and for comparison the equation of state for hydrogen matter is also shown. The numerical values of the helium matter equation of state are shown in Table 2-1, where the pressure is computed from (2.22) with $m=m_e$, the electron mass, and t deduced from (2.20) for $\rho = (m_n+m_p+m_e)n$. The kinetic energy density of the electrons may be neglected. The numerical values of the equation of state are given for equal intervals of $\log_{10}\rho$. Straight line interpolations of $\log_{10}P$ vs. $\log_{10}\rho$ between intervals should be accurate enough for stellar structure computations.

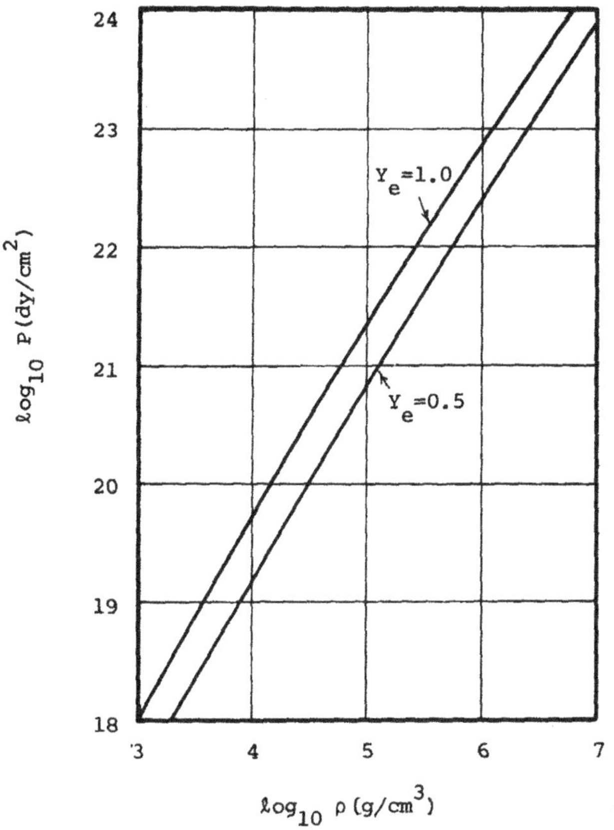

Figure 2-1. Comparison of equation of state for matter composed of helium nuclei (Y_e=0.5) with that for matter composed of hydrogen nuclei (Y_e=1.0).

Table 2-1. Numerical values of equation of state of matter composed of helium nuclei at zero temperature

t	$\rho_{He}(g/cm^3)$	$P(dy/cm^2)$	$\ell og_{10}\,\rho$	$\ell og_{10}\,P$	Γ
0.117	3.16E3	2.11E18	3.5	18.324	1.66
0.172	1.00E4	1.43E19	4.0	19.156	1.66
0.252	3.16E4	9.63E19	4.5	19.984	1.65
0.371	1.00E5	6.41E20	5.0	20.807	1.64
0.544	3.10E5	4.16E21	5.5	21.619	1.61
0.798	1.00E6	2.59E22	6.0	22.414	1.57
1.172	3.16E6	1.52E23	6.5	23.182	1.51
1.720	1.00E7	8.36E23	7.0	23.922	1.45
2.524	3.16E7	4.31E24	7.5	24.634	1.40
3.706	1.00E8	2.13E25	8.0	25.328	1.37

We note that when the Fermi momentum of the electron gas is in the non-relativistic domain, so that $t \ll 1$, the equation of state behaves as:

$$P \propto \rho^{5/3} , \tag{2.27}$$

and when the Fermi momentum turns extreme relativistic, $t \gg 1$, the equation of state behaves as:

$$P \propto \rho^{4/3} . \tag{2.28}$$

In general, the equation of state may be expressed qualitatively as:

$$P \propto \rho^{\Gamma} \tag{2.29}$$

where

$$\Gamma = \frac{d(\ell n\ P)}{d(\ell n\ \rho)} , \tag{2.30}$$

is called the <u>adiabatic index</u>. The magnitude of the adiabatic index reflects the "stiffness" of the equation of state and is crucial to the stability of stellar structure. For matter composed of helium nuclei, it varies monotonically with density as shown in Table 2-1.

Structure of White Dwarf Stars

The structure of the white dwarf stars may be studied approximately by treating it as a spherically symmetric fluid in hydrostatic equilibrium satisfying the following condition;

$$\frac{dP}{dr} = -\rho(r)\frac{GM(r)}{r^2} , \qquad (2.31)$$

with

$$M(r) = \int_0^r \rho(r')4\pi r'^2 dr' , \qquad (2.32)$$

where r is the radial distance measured from the center of the star, and G is the gravitational constant. With the prescribed relation between P and ρ given by the equation of state, (2.31) may be integrated numerically by first choosing a central density ρ_c. The corresponding central pressure is P_c, from which the pressure decreases progressively outwards layer by layer according to (2.31). The pressure at each layer gives the corresponding density for that layer with which the cumulative mass M(r) may be evaluated. This procedure is continued until a specific surface density is reached, at which point the total mass of the star and the radius of the star are thus established for that central density ρ_c. After attempting the same evaluations for a sequence of central densities, the relationship between stellar mass to central densities is established. In Figure 2-2 we show a plot of such relationship

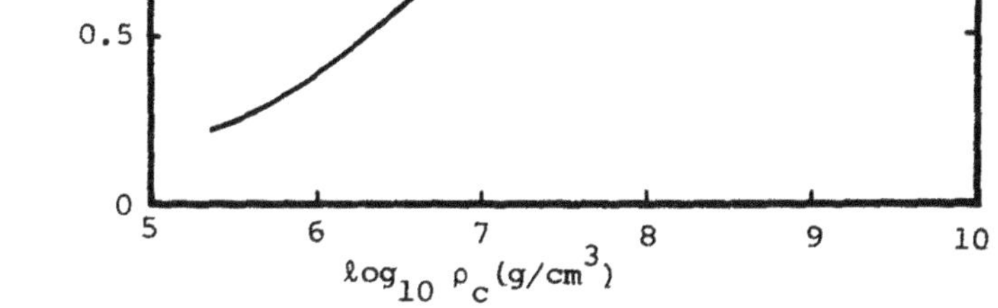

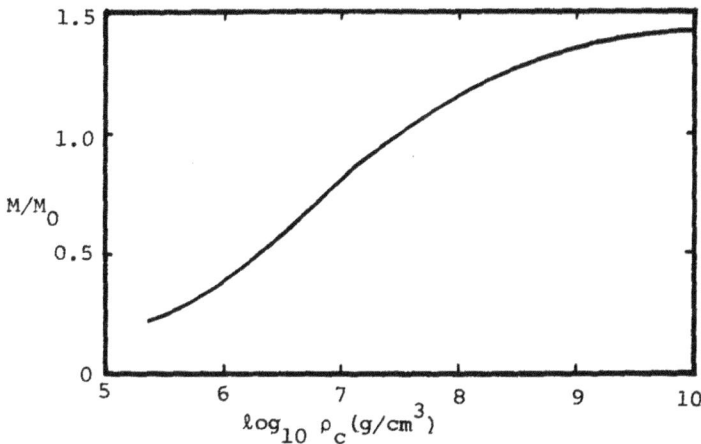

Figure 2-2. White dwarf mass, M, in units of solar mass $M_\odot$, as determined from equation of state for a completely degenerate helium gas, plotted as a function of the central density ρ_c.

obtained by Chandrasekhar (1939). The maximum mass is 1.44 $M_\odot$ at an infinite central density. However, since neutronization processes, which we shall discuss in Chapter 2, set in at approximately 10^8 g/cm^3, at which point the composition of matter begins to transform and the equation of state beyond 10^8 g/cm^3 will not support pressure as effectively as helium matter. Hence, to be realistic the maximum white dwarf mass should be that corresponding to a central density of 10^8 g/cm^3 and is given by 1.2 $M_\odot$, which we shall refer to as the <u>Chandrasekhar limit</u>.

It has been shown by Chiu (1964) that for a star to be stable under perturbations the following integral expression should be satisfied:

$$\int_0^R P(r)\,\{\Gamma(r) - \frac{4}{3}\}\, r^2 dr \;>\; 0 \;, \qquad (2.33)$$

where both the pressure P and adiabatic index Γ are functions of r. (2.33) is derived under the assumption that whenever the star experiences a contraction it contracts uniformly over its entire volume. This is a reasonable assumption. (2.33) then states that in order for the star to be stable against perturbations, the adiabatic index of the matter composing the star should be greater than 4/3 over a large portion of the star. This is an interesting result to keep in mind.

In this section, we derive the equation of state with complete neglect of electrostatic interactions. We shall come to an examination of the role of electrostatic interactions in the next section.

References

Chandrasekhar, S. (1939). An Introduction to the Study of Stellar Structure, Dover, New York.

Chiu, H.Y. (1964). In Lectures in Theoretical Physics, (Ed. W. Brittin) Vol. 6, p.225, University of Colorado Press, Boulder, Colorado.

Bibliography

Gasiorowicz, S. (1974). Quantum Physics, Wiley, New York.

Schwarzschild, M. (1958). Structure and Evolution of the Stars, Dover, New York.

3. Electrostatic Interactions

The energy density and pressure expressions derived for an ideal degenerate electron gas in the last section may be considered to be a first order estimate of such quantities in a dense plasma of electrons and atomic nuclei. The nuclei being much more massive than the electrons may be neglected for their part in the kinetic energy and pressure. They only contribute to the mass density of the system. They also interact with the electrons. We shall examine now the role of electrostatic interaction on the energy density and pressure of the electrons.

In a plasma of electrons and atomic nuclei, the electrons experience both attractive and repulsive interactions. In the first approximation these interactions are mutually suppressing and the approximation of a free electron gas is justified. This is especially true when the electron density is high and the degeneracy energy of the electron gas is correspondingly high. However, for low densities the electrostatic interactions could introduce a comparatively large correction to the free electron gas model.

The principal correction is due to the fact that the positive charges are not uniformly distributed in space but are concentrated in individual nuclei. The nuclei will organize themselves into a dense packing lattice. For whatever form of the lattice a polygonal cell of space surrounding each nucleus may be delineated. It is called a Wigner-Seitz cell in Solid State Physics. We shall replace the polygonal cell by a spherical cell of the same volume. There will be no net force acting along the periphery of the cell, which may be treated as an isolated unit with the space beyond the cell considered empty.

Let the atomic number of the nucleus be A and charge Z, the radius r_s of such a spherical cell in a system of mass density ρ is given by:

$$\text{(volume of each cell)} = \left(\frac{\rho}{Am_p}\right)^{-1} = \frac{4\pi}{3} r_s^3 = \frac{Z}{n}, \qquad (3.1)$$

where m_p is the mass of a proton, Am_p the mass of the nucleus, and n the electron number density:

$$n = \frac{3\,Z}{4\pi\, r_s^3} \qquad (3.2)$$

Let us first estimate the electron electrostatic interaction within such a cell by assuming all Z electrons are uniformly distributed. We compute separately the electron-nucleus and electron-electron interactions. We shall use the Gaussian (cgs) units of electric charge, $\alpha = e^2/\hbar c = (137.04)^{-1}$.

(1) interaction energy E_{N-e} due to electron-nucleus interaction:

$$E_{N-e} = \int_0^{r_s} \frac{Ze}{r}(-en)4\pi r^2 dr = -\frac{3}{2}\frac{(Ze)^2}{r_s} . \tag{3.3}$$

(2) interaction energy E_{e-e} among Z electrons:

First, the electric field for a uniformly charge sphere is given by:

$$\begin{aligned}(\text{electric field}) &= -\frac{Ze}{r^2} , \quad (r \geq r_s) \\ &= -\frac{Zer}{r_s^3} , \quad (r < r_s) .\end{aligned}$$

The electric potential inside the cell is therefore:

$$\Phi(r) = \int_\infty^{r_s} \frac{Ze}{r^2} dr + \int_{r_s}^{r} \frac{Zerdr}{r_s^3} = -\frac{Ze}{r_s}\left(\frac{3}{2} - \frac{1}{2}\frac{r^2}{r_s^2}\right) .$$

Then, the interaction energy is:

$$E_{e-e} = \frac{1}{2}\int_0^{r_s} (-en)\Phi(r)4\pi r^2 dr = \frac{3}{5}\frac{(Ze)^2}{r_s} . \tag{3.4}$$

(3) the total electrostatic energy of each cell is:

$$E = E_{N-e} + E_{e-e} = -\frac{9}{10}\frac{(Ze)^2}{r_s} . \tag{3.5}$$

Substituting for r_s from (3.2), the electrostatic energy per electron in the entire system is:

$$\varepsilon_e = \frac{E}{Z} = -\frac{9}{10}\left(\frac{4\pi}{3}\right)^{1/3} Z^{2/3} e^2 n^{1/3} . \tag{3.6}$$

The interaction energy is negative and thus contributes to a reduction in the pressure of the system. However, the assumption that electrons are uniformly distributed within the cell is certainly unjustified. The electron distribution is greatly influenced by the charge of the nucleus which tends to attract electrons towards it creating a denser distribution of electrons in it neighborhood. Since a complete determination of the electron distribution is difficult, one can estimate the variation in the distribution in a semi-classical way. This is called the Thomas-Fermi method.

The Thomas-Fermi Method

The Hartree equations of (1.15) provide the conditions for the determination of the Hartree single-particle energies ε_i^H. There ϕ_i denote the electron wave functions within a cell. The electrons interact with the nucleus through $V(\vec{x}_i)$ and among themselves through $v(\vec{x}_i-\vec{x}_j)$. There will be N=Z number of electrons within the cell. After the Hartree equations are solved self-consistently and all the single-electron wave functions are found, the distribution of electrons within the cell is determined. Instead of embarking on a solution of the Hartree equations, the Thomas-Fermi method develops its steps on the assumption that all single-electron wave functions may be approximated by plane waves as in a uniform system. This approach would be accurate if the potential experienced by the electrons is sufficiently uniform within the cell. In any case, the method would give a reasonable first order correction to the assumption of uniformity.

For a particular plane wave with momentum $\vec{p}$, the Hartree equation (1.15) is replaced by a classical expression for single-electron energy:

$$\varepsilon = \frac{p^2}{2m} - e\Phi \quad ,$$

where Φ is the electrostatic potential determined by the Gauss' law:

$$\nabla^2\Phi = \frac{1}{r^2}\frac{d}{dr}\left[r^2\frac{d}{dr}\Phi\right] = -4\pi\left\{Ze\delta(r) - en(r)\right\} , \tag{3.7}$$

where n(r) is the electron density distribution. In the Hartree method n(r) is given by:

$$n(r) = \sum_{\substack{j=1 \\ j\neq i}}^{Z} \phi_j^*(\vec{r})\phi_j(\vec{r}) \tag{3.8}$$

where n(r) may be assumed to be spherically symmetric and is a function of the radial distance r. In the Thomas-Fermi method, n(r) is approximated by:

$$n(r) = \frac{8\pi}{3h^3}\, p_F^3(r) \quad , \tag{3.9}$$

which imitates the result (2.7) of a translationally invariant system. Here the variation of n as a function of r is anticipated by writing the Fermi momentum p_F a function of r. The dependence of p_F on r is to be determined by requiring the maximum energy, or the Fermi energy ε_F, of the electron at every point in the cell to be the same. This is a general result from statistical mechanics for a system in thermal equilibrium. In the language of statistical mechanics such a maximum energy is denoted by μ, the chemical potential, which we shall also use here:

$$\mu = \varepsilon_F = \frac{p_F^2(r)}{2m} - e\,\Phi(r) \; . \tag{3.10}$$

Thus, $p_F(r)$ and $\Phi(r)$ are related by a constant μ which is determined by the charge number Z, and $p_F(r)$ may be expressed in terms of $\Phi(r)$. With this substitution (3.8) is converted into a nonlinear differential equation for $\Phi(r)$, which for $r > 0$ is given by:

$$\frac{1}{r^2}\frac{d}{dr}\left(r^2\frac{d\Phi}{dr}\right) = 4\pi\left(\frac{8\pi e}{3h^3}\right)\{2m(\mu+e\Phi)\}^{3/2} \; . \tag{3.11}$$

It can be simplified by introducing the following variables. Let us write:

$$(\mu+e\phi) = Ze^2\left(\frac{u}{r}\right),$$

and

$$r = ax \; ,$$

where

$$a = \left(\frac{9\pi^2}{128Z}\right)^{1/3} a_0 \; ,$$

with $a_0 = (\hbar/mc)(\hbar c/e^2)$ = Bohr radius. Then (3.11) reads:

$$\frac{d^2u}{dx^2} = \frac{u^{3/2}}{x^{1/2}} \; . \tag{3.12}$$

Solutions of (3.11) or (3.12) would have to satisfy the following boundary conditions:

(1) $r \to 0$: $\Phi \to \frac{Ze}{r}$, which is the potential due to the nucleus alone.
In the new variables this is expressed as: $u \to 1$.

(2) $r = r_s$: $\frac{d\Phi}{dr} = 0$, since the nucleus is completely shielded by the surrounding electrons and the electric field is zero at the boundary of the cell. This means: $\frac{du}{dr} = \frac{u}{r}$.

In summary, u is determined by the following set of equations:

$$\frac{d^2u}{dx^2} = \frac{u^{3/2}}{x^{1/2}} ,$$

$$x = 0: \quad u = 1,$$

$$x = x_s: \quad \frac{du}{dx} = \frac{u}{x} . \tag{3.13}$$

Analytic solution of u is difficult to obtain due to the nonlinearity of the differential equation. Consequently, the Thomas-Fermi method is usually associated with numerical solutions.

A straight forward numerical integration of (3.13) from x=0 outwards also encounters difficulty, since the differential equation is singular there. It is therefore necessary to initiate the numerical integration at some point x away from the origin. In order to accomplish this, let us make an expansion of u in a power series of the form:

$$u = 1 + a_2x^2 + a_3x^{3/2} + a_4x^2 + \ldots + a_nx^{n/2} + \ldots \tag{3.14}$$

(3.13) then relates all expansion coefficients a_n to a_2. Hence, once a_2 is chosen, u may be evaluated at some point x, within the radius of convergence of the series, to arbitrary accuracy by terminating the series at a specific number of terms. The slope of u is obtained by differentiating the power series and its value at the same point may similarly be evaluated. The numerical integration of u from that point on may then ensue. The value of u at x_s is then checked for the requirement of the boundary conditions. A properly chosen a_2 will enable a solution u to be found. The expansion coefficients are related to a_2 as follows:

a_2 = initial slope
$a_3 = (4/3)$
$a_4 = 0$
$a_5 = (2/5)a_2$
$a_6 = (1/3)$
$a_7 = (3/70)a_2^2$
$a_8 = (2/15)a_2$
$a_9 = (2/27) - (1/252)a_2^3$
$a_{10} = (1/175)a_2^2$
$a_{11} = (31/1485)a_2 + (1/1056)a_2^4$
. . . .

Feynman, Metropolis, Teller (1949), who had carried out such a numerical method, suggested the following procedure;

(1) introduce a change of independent variable: $x = (\omega^2/2)$, which makes the interval for each step of numerical integration conveniently small near the origin where u changes appreciably, and automatically increases the interval further out where the function changes more slowly.

(2) the starting point of the numerical integration is chosen to be $\omega=0.9$, and u is evaluated from the power series (3.15) at that point after choosing a value for a_2.

(3) the slope of u at the same point is found from evaluating u by the same method at $\omega=0.88$ and $\omega=0.92$, and then divide the difference in u by 0.04. They found this gives a more accurate evaluation of the slope than that given by a differentiated series.

(4) intervals of numerical integration are taken to be $\omega=0.04$; the error in each step is kept $<0.00002_5$ in u.

Some of the results are as follows: for $a_2=-1.58808$, numerical integration shows that $\frac{du}{dx}=0$ at $x=\infty$, which is the boundary condition for $x_s=\infty$. This is the case of a single atom. For a densely packed situation, it must be that $a_2>-1.58808$, for then $\frac{du}{dx}=\frac{u}{x}$ is satisfied at a finite x. For example, with $a_2=-1.58806$, the boundary condition is satisfied at $x_s=7.385$ with the value of $u(x_s)=0.0979$ there.

In terms of u and x, the Fermi momentum within the cell is given by:

$$p_F = \left[2m\left(\frac{Ze^2}{a}\right)\left(\frac{u}{x}\right)\right]^{1/2} = \frac{\hbar}{a_0}\left(\frac{128Z^4}{9\pi^2}\right)^{1/6}\sqrt{2}\left(\frac{u}{x}\right)^{1/2}, \tag{3.15}$$

and the electron density:

$$n_{TF}=\frac{8\pi}{3h^3}p_F^3 = \frac{32Z^2}{9\pi^3 a_0^3}\left(\frac{u}{x}\right)^{3/2}. \tag{3.16}$$

A typical case of the electron distribution within a cell as determined bv the Thomas-Fermi method is shown in Figure 3-1. It shows a distribution very different from the uniform distribution. A better comparison is to plot $r^2n(r)$ for the two cases, and this is shown in (b).

The total charge Q within the cell can be found from a direct integration of n:

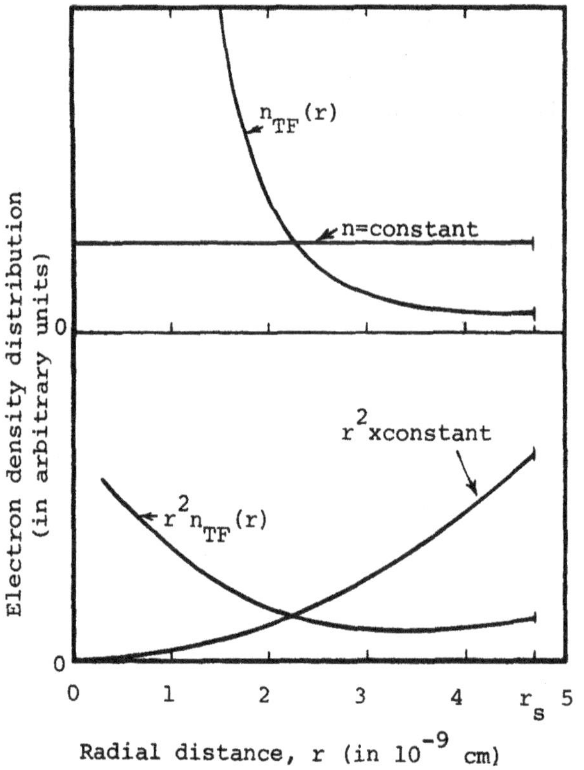

Figure 3-1. Electron density distribution given by the Thomas-Fermi method.

$$Q = 4\pi \int_0^{r_s} r^2 dr\, n(r) = \frac{8\pi}{3h^3}\left(2m\,\frac{Ze^2}{a}\right)^{3/2} 4\pi a^3 \int_0^{x_s} x^2 dx \left(\frac{u}{x}\right)^{3/2}$$

$$= Z \int_0^{x_s} x dx\, \frac{d^2u}{dx^2} = Z\int_0^{x_s} dx \frac{d}{dx}\left(x\, \frac{du}{dx} - u\right)$$

$$= Z\, \{x_s \left.\frac{du}{dx}\right|_{x_s} - u(x_s) + u(0)\} = Z\ ,$$

where the use is made of the differential equation and the boundary conditions for u.

The pressure from a system of cells of the type described here is the same as that due to a system of free electrons with a density equal to the electron density at the boundary of the cell where no net force acts. Writing the same pressure expression as (2.10),

$$P = \frac{2}{3h^3} \int_0^{p_F(r_s)} 4\pi p^2 dp (pv)\ ,$$

where,

$$v = \frac{d}{dp}\left(\frac{p^2}{2m} - e\Phi\right) = \frac{p}{m} .$$

Thus, the pressure of the system from the Thomas-Fermi method is:

$$P = \frac{\alpha^2}{15\pi^2}\left(\frac{mc^2}{a_0^3}\right)\left[\left(\frac{128Z^4}{9\pi^2}\right)^{1/3} 2\frac{u(x_s)}{x_s}\right]^{5/2} \tag{3.17}$$

where $\alpha=(137.04)^{-1}$ is the fine structure constant, and $a_0=5.292\text{x}10^{-9}$ cm is the Bohr radius. The factor $(mc^2/a_0^3)=5.521\text{x}10^{18}$ dyne/cm^2 has the dimensions of pressure. The matter density is given by:

$$\rho = \frac{A\, m_p}{\frac{4\pi}{3}r_s^3} = \frac{32}{3\pi^3}\left(\frac{m_p}{a_0^3}\right)\frac{AZ}{x_s^3} \tag{3.18}$$

where A is the atomic number of the atom type forming the system.

The Thomas-Fermi method treats the interaction energies classically and ignores the fact that the antisymmetry in the total electron wave function gives rise to the exchange effects. Such effects are exhibited in the Hartree-Fock equations. In order to include the exchange effects, the Thomas-Fermi method must be modified. This was carried out by Dirac (1930). We shall refer to the modified method as the Thomas-Fermi-Dirac method.

The Thomas-Fermi-Dirac Method

The Hartree-Fock equations (1.25) differ from the Hartree equations (1.15) in two respects. First, the direct interaction terms among the electrons compose of a summation which is over all electrons (instead of Z-1 electrons). They correspond exactly to the classical potential energy function:

$$e\Phi(\vec{x}) = V(\vec{x}) + \sum_{j=1}^{Z}\int\phi_j^*(\vec{y})\frac{e^2}{|\vec{x}-\vec{y}|}\phi_j(\vec{y})\, d^3y . \tag{3.19}$$

Secondly, there are the exchange terms which are absent in the Hartree equations. The exchange terms are given by:

$$U_{ex}(\vec{x}) = e^2\sum_{j=1}^{Z}\phi_j(\vec{x})\,\delta(\sigma_j,\sigma_i)\int\frac{\phi_j^*(\vec{y})\phi_i(\vec{y})\, d^3y}{|\vec{y}-\vec{x}|} \tag{3.20}$$

U_{ex} may be evaluated approximately by plane wave states, and this is consistent with the Thomas-Fermi method approach. We therefore write:

$$\phi_i(\vec{x}) = (\Omega)^{-1/2} \exp(i \vec{p}\cdot\vec{x}/\hbar) ,$$
$$\phi_j(\vec{x}) = (\Omega)^{-1/2} \exp(i \vec{p}'\cdot\vec{x}/\hbar) , \tag{3.21}$$

and convert the summation over j into an integral over the momentum space:

$$\sum_j \delta(\sigma_j,\sigma_i) \rightarrow \frac{\Omega}{h^3} \int d^3p' \,\rho(p')$$

where

$$\rho(p') = 1 \quad \text{for} \quad p' \leq p_F$$
$$= 0 \quad \text{for} \quad p' > p_F . \tag{3.22}$$

The δ-function on spin components specifies that only electrons of the same spin component should contribute to the summation. The exchange terms are then evaluated as:

$$U_{ex}(\vec{x}) = \frac{e^2}{h^3}(\Omega)^{-1/2} \int d^3p'\rho(p')\int d^3x' \frac{\exp(-i\vec{p}'\cdot\vec{x}'/\hbar)\exp(i\vec{p}'\cdot\vec{x}/\hbar)\exp(i\vec{p}\cdot\vec{x}'/\hbar)}{|\vec{x} - \vec{x}'|}$$
$$= \frac{e^2}{h^3}(\Omega)^{-1/2}\exp(i\vec{p}\cdot\vec{x}/\hbar)\int d^3p'\rho(p')\int d^3r \frac{\exp\{-i(\vec{p}-\vec{p}')\cdot\vec{r}/\hbar\}}{r} ,$$

where $\vec{r} = \vec{x}'-\vec{x}$. The integrals can be evaluated:

$$U_{ex}(\vec{x}) = \frac{e^2}{\pi h} \phi_i(\vec{x}) \int d^3p'\rho(p') \frac{1}{|\vec{p}-\vec{p}'|^2}$$
$$= \frac{e^2}{h}\{ \frac{p_F^2-p^2}{p} \ln \left|\frac{p_F+p}{p_F-p}\right| + 2p_F\} \phi_i(\vec{x}) . \tag{3.23}$$

The single-electron energy is therefore:

$$\varepsilon = \frac{p^2}{2m} - e\Phi - \frac{e^2}{h} \left(\frac{p_F^2-p^2}{p} \ln\left|\frac{p_F+p}{p_F-p}\right| + 2p_F\right) . \tag{3.24}$$

The condition of equilibrium marked by the chemical potential μ is given by:

$$\mu = \frac{p_F^2(r)}{2m} - e\Phi(r) - \frac{2e^2}{h} p_F(r) , \tag{3.25}$$

which is obtained from (3.24) by setting $p=p_F$. Depending on the variation of $\Phi(r)$, different parts of the cell would reach different values of Fermi momentum $p_F(r)$. As in the Thomas-Fermi method (3.25) allows $p_F(r)$ to be expressed in terms of $\Phi(r)$:

$$p_F(r) = (2e^2m/h) + \{ (2e^2m/h)^2 + 2m(\mu + e\Phi)\}^{1/2} . \quad (3.26)$$

The electron density n is still given by (3.9), and the Gauss' law for Φ now reads:

$$\frac{1}{r^2}\frac{d}{dr}\left\{ r^2 \frac{d\Phi}{dr} \right\} = 4\pi\left(\frac{8\pi e}{3h^3}\right)\left\{(2e^2m/h) + \{ (2e^2m/h)^2 + 2m(\mu + e\Phi)\}^{1/2} \right\}^3 . \quad (3.27)$$

It is simplified by writing,

$$\{ 2m(e^2/h)^2 + (\mu + e\Phi)\} = Ze^2\left(\frac{w}{r}\right),$$

and

$$\kappa = (6\pi/Z^2)^{1/3} ,$$

with r=ax as in (3.12). The differential equation for w and its boundary conditions may be expressed as:

$$\frac{d^2w}{dx^2} = x\left(\kappa + \frac{w^{1/2}}{x^{1/2}}\right)^3 ,$$

$$w(0) = 1 ,$$

$$\left.\frac{dw}{dx}\right|_{x=x_s} = \left.\frac{w}{x}\right|_{x=x_s} . \quad (3.28)$$

As in the Thomas-Fermi method, the solution to (3.28) is to be evaluated numerically. The power series expansion of w valid for small x is exactly the same as it is for u in the Thomas-Fermi method:

$$w = 1 + a_2x + a_3x^{3/2} + a_4x^2 + \ldots + a_nx^{n/2} + \ldots \quad (3.29)$$

These coefficients are related to a_2 as follows:

a_2 = initial slope

$a_3 = (4/3)$

$a_4 = (3/2)\kappa$

$a_5 = (2/5)a_2 + (4/5)\kappa^2$

$a_6 = (1/3) + (1/2)a_2\kappa + (1/6)\kappa^3$

$a_7 = (6/35)a_2\kappa^2 + (3/70)a_2^2 + (5/7)\kappa$

$a_8 = (2/15)a_2 + (77/120)\kappa^2$

$a_9 = (2/27) - (1/252)a_2^3 + (11/35)a_2\kappa - (1/42)a_2^2\kappa^2 + (10/63)\kappa^3 +$
$+ (16/105)\kappa^4$

. . . . . .

For each specific type of atoms with atomic number A and charge Z and for a certain cell size x_s, the solution w may be determined numerically in the same manner described for the Thomas-Fermi method.

Numerical solutions of the Thomas-Fermi-Dirac method have been given by Slater and Krutter (1935) for Z=3,11,29, by Jensen (1938) for A=18,36,54 and by Feynman, Metropolis, and Teller (1949) for A=6,92. Feynman et al. compiled all data to make possible interpolation for any Z value. The results can be summarized by expressing the pressure in the following form:

$$P = f(\xi)\, P_{free}\,, \tag{3.30}$$

where

$$P_{free} = \frac{h^2}{5m}\left(\frac{3}{8\pi}\right)^{2/3}\left(\frac{Z\rho}{Am_p}\right)^{5/3} \tag{3.31}$$

is the pressure from an ideal degenerate electron gas given by (2.16), and,

$$f(\xi) = \frac{1}{3}x_s^3\left[\kappa + \left(\frac{w(x_s)}{x_s}\right)^{1/2}\right]^3\left[1 - \frac{5/4}{\kappa+\left(\frac{w(x_s)}{x_s}\right)^{1/2}}\right]^{3/5} \tag{3.32}$$

with ξ defined to be:

$$\xi = \frac{a_0}{Z^{2/3}}\left(\frac{3Z}{4\pi a x_s}\right)^{1/3} = \frac{0.701}{x_s}\,.$$

Plots of $f(\xi)$ for different Z values are shown by Feynman et al. We reproduce here in Figure 3-2 the interpolated plot of $f(\xi)$ for Z=3 and 26, which represent roughly the range of atomic types for matter below a density of 10^8 g/cm^3 that is of astrophysical interest. Add for comparison is the case without exchange interaction which is the result of the Thomas-Fermi method. As it is apparent from Figure 3-2 the pressure

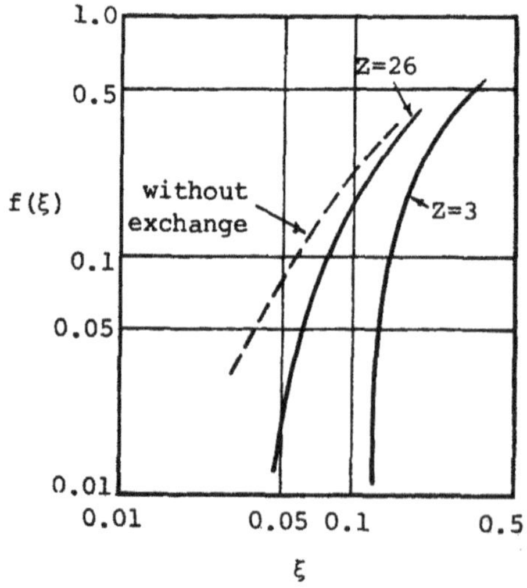

Figure 3-2. The $f(\xi)$ of (3.32) determined from numerical solutions of the Thomas-Fermi-Dirac method.

deduced from Thomas-Fermi-Dirac method is always below that of the ideal degenerate electron gas. There is a larger reduction in pressure for low Z material than it is for high Z. Also, the exchange interaction introduces further reduction than the case without it, and hence the Thomas-Fermi-Dirac pressure is lower than the Thomas-Fermi pressure.

As matter density increases the radius of individual cells will decrease. A reduction in r_s means an increase in ξ and in $F(\xi)$, which tends more and more towards unity, with the pressure $P \to P_{free}$. Suppose we say that $F(\xi) \simeq 1$ for $\xi \simeq 1$, then we can determine an approximate density over which a free degenerate electron gas approximation should be valid. For $\xi = 1$,

$$r_s = \left(\frac{9\pi^2}{128}\right)^{1/3} \frac{a_0}{Z^{1/3}}(0.701) \tag{3.33}$$

which yields a matter density of,

$$\rho_1 = \frac{Am_p}{(4\pi/3)r_s^3} \simeq 10AZ \text{ g/cm}^3 \tag{3.34}$$

Thus, for Fe^{56}, $\rho_1 \simeq 10^4$ g/cm^3, above which the ideal electron gas approximation may be considered valid. Below this density the Thomas-Fermi-Dirac correction should be imposed, but the method is in general not suitable for matter densities which are too low largely because the orbital motions of the electrons about the nucleus have not been properly taken into account. For example, the results obtained here would not be applicable to metallic iron.

On the basis of the work of Feynman et al., Harrison, Thorne, Wakano and Wheeler (1965) have given an analytic expression of the equation of state of matter composing of iron atoms up to a density of $\rho \simeq 10^6$ g/cm^3:

$$P = 3.0271\text{x}10^{12} (\rho^{1/3} - 1.4415)^5 - 1.50319\text{x}10^{11} \tag{3.35}$$

where ρ in g/cm^3 and P in dyne/cm^2.

An Approximate Coulomb Exchange Energy

The exchange energy term in the Hartree-Fock equation complicates the solution of the equations enormously. Approximate estimates of such exchange terms may be employed whenever they do not play a dominant role in the computation. Such occasions arise in the nuclear problem when the Coulomb energy plays only a perturbative role. An approximate Coulomb exchange energy $E_{ex,C}$ is given by Slater (1951, 1960) obtained by integrating the exchange energy given in (3.24) over the filled states:

$$\begin{aligned} E_{ex,C} &= -\frac{e^2}{h}\left(\frac{\Omega}{h^3}\right)\int_0^{p_F} 4\pi p^2 dp\left\{\frac{p_F^2-p^2}{p}\,\ell n\left|\frac{p_F+p}{p_F-p}\right| + 2\,p_F\right\} \\ &= \frac{e^2\Omega}{h^4}\,4\pi p_F^4 = \frac{3e^2\Omega}{4}\left(\frac{3}{\pi}\right)^{1/3} n^{4/3}, \end{aligned} \tag{3.36}$$

where n is the electron number density, $n = (8\pi p_F^3)/(3h^3)$. This expression will be useful later.

References

Dirac, R.A.M. (1930). Proc. Camb. Phil. Soc. 26, 376.

Feynman, R.P., Metropolis, N. and Teller, E. (1949). Phys. Rev. 75, 1561.

Harrison, B.K., Thorne, K.S., Wakano, M. and Wheeler, J.A. (1965). Gravitation Theory and Gravitational Collapse, Chicago Univ. Press Chicago.

Jensen, H. (1938). Zeit. f. Physik 111, 373.

Slater, J.C. and Krutter, H.M. (1935). Phys. Rev. 47, 559.

Slater, J.C. (1951). Phys. Rev. 81, 385.

Slater, J.C. (1960). Quantum Theory of Atomic Structure, Vol. 2, McGraw-Hill, New York.

Bibliography

Fermi, E. (1928). Z. Physik 48, 73.

Hartree, D.R. (1928). Proc. Camb. Phil. Soc. 24, 111.

Thomas, L.H. (1927). Proc. Camb. Phil. Soc. 23, 542.

Zeldovich, Ya. B. and Novikov, I.D. (1971). Relativistic Astrophysics, Chicago Univ. Press, Chicago.

4. Thermal Effects

In this section we shall develop the general formalism for dealing with thermal effects, which like particle interactions can add greatly to the complexity of the many-body problem. We shall treat thermal effects only to an accuracy which is consistent with the Hartree-Fock approximation of particle interactions. The following is a brief summary of results from thermodynamics and statistical mechanics.

In the previous sections we have been concentrating on establishing the ground state of a many-body system. There will be excited states for the system with energies above the ground state energy. Let us consider all eigenstates of the many-body Hamiltonian H and label them by the subscript n: $H\,\Psi_n = E_n\Psi_n$. Each eigenstate represents a stationary state of the system. In statistical mechanics we envision that the system exists within a much larger system (or a "heat bath"),so that it may draw energy freely from the larger system (but with negligible interaction with it) and may therefore exist in any one of its stationary states. The totality of states of this N-body system forms an ensemble, called the canonical ensemble. From statistical mechanics it is determined that at a temperature T the relative probability for the N-body system to be in any one of its states Ψ_n is given by the Boltzmann factor $\exp(-\beta E_n)$, where $\beta = (k_B T)^{-1}$. In other words, the density matrix in the canonical ensemble is weighted by the Boltzmann factor. The partition function in the canonical ensemble is given by:

$$Q_N = \sum_n e^{-\beta E_n}\,, \tag{4.1}$$

where the sum is taken over all stationary states of the system which has its particle number restricted to N in the volume Ω. All thermodynamical quantities can in principle be deduced from Q_N once it is known. However,the evaluation of Q_N would be impossible if each state Ψ_n requires a separate computation of the type described before for the ground state case. For the method to be useful Q_N would have to be expressible in an analytic form. To accomplish this approximate method have been devised. There are a variety of them, and their usefulness depends on the nature of the problem under consideration. We shall discuss one method which is related to the Hartree-Fock approximation but before we come to this let us review the situation related to the case of an ideal Fermi gas, for which the problem is completely solvable.

For an ideal Fermi gas the single-particle states may be determined individually, independent of each other and of N. Labelling the single-particle states by the subscript i, and introducing a set of occupation numbers n_i, where n_i=1 or 0, specifying whether or not the ith single-particle state is occupied, the total energy of the system with occupation numbers given by the set $\{n_i\}$ may be expressed as:

$$E_n = \sum_i \varepsilon_i n_i \equiv E\{n_i\} , \tag{4.2}$$

with

$$\varepsilon_i = p_i^2/(2m) , \tag{4.3}$$

and

$$N = \sum_i n_i . \tag{4.4}$$

The partition function (4.1) may be replaced by:

$$Q_N = \sum_{\{n_i\}} e^{-\beta E\{n_i\}} , \tag{4.5}$$

where the summation is over all possible sets of occupation numbers $\{n_i\}$ subject to the restriction (4.4).

Q_n is constructed for an ensemble of states of a system whose particle number is restricted to N. Because of this restriction, Q_N is still difficult to evaluate even for the ideal gas. Thus, the concept of a canonical ensemble is generalized to that of a grand canonical ensemble, for which the system is allowed to exchange not only energy but also particles with the surrounding heat bath. The grand canonical ensemble is composed of stationary states of not just one system of fixed particle number N but systems of the same volume Ω with arbitrary numbers of particles. The partition function in the grand canonical ensemble is called the grand partition function, and is given by:

$$Z = \sum_{N=1}^{\infty} z^N Q_N , \tag{4.6}$$

where $z = \exp(\beta\mu)$ is called the fugacity, and μ, the chemical potential. For an ideal Fermi gas, Z can be evaluated readily:

$$Z = \sum_{N=0}^{\infty} \sum_{\substack{\{n_i\} \\ \Sigma n_i = N}} z^{\Sigma n_i} e^{-\beta \Sigma \varepsilon_i n_i}$$

$$= \sum_{n_0} \sum_{n_1} \cdots \sum_{n_\infty} (ze^{-\beta\varepsilon_0})^{n_0} (ze^{-\beta\varepsilon_1})^{n_1} \cdots\cdots$$

$$= \{ 1 + ze^{-\beta\varepsilon_0}\} \{ 1 + ze^{-\beta\varepsilon_1}\} \cdots\cdots$$

$$= \prod_{i=0}^{\infty} \{ 1 + ze^{-\beta\varepsilon_i}\} \ . \qquad (4.7)$$

For a specific value of the fugacity z, or equivalently the chemical potential μ, the most probable number of particles N in the system is given by the ensemble average:

$$N = Z^{-1} \sum_{N=0}^{\infty} N z^N Q_N = z \frac{\partial}{\partial z} (\ln Z) \ . \qquad (4.8)$$

The most probable occupation number in each state is:

$$\bar{n}_i = Z^{-1} \sum_{N=0}^{\infty} z^N \sum_{\{n_i\}} n_i e^{-\beta E} = - \beta^{-1} \frac{\partial}{\partial \varepsilon_i} (\ln Z) \ , \qquad (4.9)$$

and the most probable energy of the system:

$$E = Z^{-1} \sum_{N=0}^{\infty} z^N \sum_{\{n_i\}} E e^{-\beta E} = - \frac{\partial}{\partial \beta} (\ln Z) \ . \qquad (4.10)$$

From these relations we derive for an ideal Fermi system the following results:

(1) $$\bar{n}_i = \{ 1 + e^{\beta(\varepsilon_i - \mu)}\}^{-1} \ , \qquad (4.11)$$

(2) $$N = \sum_i \bar{n}_i \ , \qquad (4.12)$$

(3) $$E = \sum_i \varepsilon_i \bar{n}_i \ . \qquad (4.13)$$

The equation of state is a functional relationship among the thermodynamic parameters for a system in thermal equilibrium. It can be deduced from the grand partition function by identifying:

$$P\Omega = \beta^{-1}(\ln Z) = \beta^{-1} \sum_i \ln(1 + z\, e^{-\beta\varepsilon_i}) . \tag{4.14}$$

The entropy S of a system in thermal equilibrium is defined in statistical mechanics to be;

$$S = k_B \ln N_Q , \tag{4.15}$$

where N_Q is the number of quantum states accessible to the system. For a Fermi system in a completely degenerate state, $N_Q = 1$, i.e., at $T = 0$ the system is permitted a single quantum state, the ground state, and consequently the entropy is zero. At finite temperatures, N_Q may be deduced from combinatorics. When the particles in the system are completely uncorrelated as in the case of an ideal Fermi gas, the entropy is found to be:

$$S/k_B = -\sum_i \{\bar{n}_i \ln \bar{n}_i + (1 - \bar{n}_i) \ln(1 - \bar{n}_i)\} , \tag{4.16}$$

which again evaluates to $S = 0$ for a completely degenerate system where $\bar{n}_i = 1$ or 0.

For a uniform system the summation over single-particle states in the above expressions may be replaced by integral over the momentum space as before:

$$\sum_i \rightarrow \gamma(\Omega/h^3) \int d^3p , \tag{4.17}$$

where γ is the degeneracy factor for each state ($\gamma=2$ for electrons). In summary, the various thermodynamic parameters are given by:

(1) particle number density,

$$n = \frac{N}{\Omega} = (\gamma/h^3)\int d^3p\, \bar{n}_p , \tag{4.18}$$

(2) energy density,

$$\varepsilon = \frac{E}{\Omega} = (\gamma/h^3)\int d^3p\, \varepsilon_p\, \bar{n}_p , \tag{4.19}$$

(3) pressure,

$$\begin{aligned} P &= \beta^{-1}(\gamma/h^3)\int d^3p \ln\{1 + \exp(-\beta(\varepsilon_p - \mu))\} \\ &= (\gamma/h^3)\int d^3p\, \frac{1}{3}\, p\left(\frac{\partial\varepsilon_p}{\partial p}\right) \bar{n}_p , \end{aligned} \tag{4.20}$$

(4) entropy density,

$$S/(k_B\Omega) = -(\gamma/h^3)\int d^3p\ \{\ \bar{n}_p\ \ell n\ \bar{n}_p + (1-\bar{n}_p)\ell n(1-\bar{n}_p)\}$$
$$= (\beta\gamma/h^3)\int d^3p\ \{\varepsilon_p + \frac{1}{3}p\frac{\partial\varepsilon_p}{\partial p}\}\ \bar{n}_p - \beta\mu n\ . \quad (4.21)$$

where $\varepsilon_p = p^2/2m$,

and

$$\bar{n}_p = \{1 + \exp\big(\beta(\varepsilon_p-\mu)\big)\}^{-1}\ . \quad (4.22)$$

The last lines in (4.20) and (4.21) are obtained after integrations by parts are performed. The expression for pressure thus obtained is the same as that before in (2.13). (4.21) expresses one of the known thermodynmaic relations written usually in the form.

$$E(S,\Omega,N) = TS - P\Omega + \mu N\ , \quad (4.23)$$

where E is a function of the independent variables S, Ω, and N, while:

$$T = \left(\frac{\partial E}{\partial S}\right)_{\Omega,N}\ , \quad -P = \left(\frac{\partial E}{\partial \Omega}\right)_{S,N}\ , \quad \mu = \left(\frac{\partial E}{\partial N}\right)_{S,\Omega}\ . \quad (4.24)$$

Thermodynamic functions based on other independent variables may be obtained from E by means of the Legendre transformations. They are:

Helmholtz free energy: $F(T,\Omega,N) = E(S,\Omega,N) - TS\ ,$ (4.25)

Gibbs free energy: $G(T,P,N) = F(T,\Omega,N) + P\Omega\ ,$ (4.26)

Thermodynamic potential: $\Phi(T,\Omega,\mu) = F(T,\Omega,N) - \mu N\ .$ (4.27)

The independent variables of these functions are explicitly specified. Making use of (4.23), we have:

$$F = -P\Omega + \mu N\ , \quad G = \mu N\ , \quad \Phi = -P\Omega\ . \quad (4.28)$$

From (4.28) we see that the chemical potential in a one-component system is just the Gibbs free energy per particle:

$$\mu(T,P) = G(T,P,N)/N\ , \quad (4.29)$$

and the pressure of the system is:

$$P(T,\mu) = -\Phi(T,\Omega,\mu)/\ \Omega. \quad (4.30)$$

Note that μ is no longer dependent on N, or P on Ω.

A system in thermal equilibrium may not be in a single uniform state, but falls into two uniform parts in contact with each other.

These two parts are said to be in two different <u>phases</u> of the system. To be in thermal equilibrium, the temperatures of the two phases must be equal. In addition, the pressures and the chemical potentials of the two phases must also be equal. All of these conditions can be derived from the requirement that the entropy of the system is at a maximum. Since the chemical potentials are functions of P and T, the equality:

$$\mu_a(P,T) = \mu_b(P,T) \ , \tag{4.31}$$

implies a relation between P and T at phase equilibrium. Conditions of phase equilibrium will be examine further in Section 10.

The above situation is to be distinguished from that of a mixture of two gases which do not transform into each other. In a mixture, the total pressure of the system is just the sum of the partial pressures.

In order to see the implications of the thermal effects on the equation of state of an ideal Fermi gas, we shall make explicit evaluations of n, ε, P and s from (4.18) to (4.21) with μ as an implicit parameter. We shall do this for a free neutron gas, for which non-relativistic treatment is adequate for k_BT in the range of several MeV. Relativistic generalization of these expressions may be carried out in the same manner as it is done in Section 2 and will not be given here. The standard low temperature ($k_BT<<\mu$) expansion of $\bar{n}_p$ is given by:

$$\bar{n}_p = \Theta(\mu - \varepsilon_p) - (\pi^2/6)(k_BT)^2 \frac{\partial}{\partial\varepsilon_p}\delta(\varepsilon_p - \mu) + \ldots . \tag{4.32}$$

where Θ is the step-function:

$$\Theta(x) = \begin{cases} 1 & x \geq 0 \\ 0 & x < 0 \ . \end{cases} \tag{4.33}$$

We shall not employ approximation (4.32), since μ may not be large compared to k_BT. All of the thermodynamic parameters may be expressed in terms of Fermi integrals, defined to be:

$$F_t(\eta) = \int_0^\infty \frac{x^t\,dx}{1+\exp(x-\eta)} \ . \tag{4.34}$$

Writing $\beta\mu = \eta$, we have:

$$\text{(a)} \quad n = (4\pi/h^3)(2mk_BT)^{3/2} F_{1/2}(\eta) , \tag{4.35}$$

$$\text{(b)} \quad \varepsilon = \frac{3}{2} P = (4\pi/h^3)(2mk_BT)^{3/2}(k_BT) F_{3/2}(\eta) , \tag{4.36}$$

$$\text{(c)} \quad S/k_B\Omega = (4\pi/h^3)(2mk_BT)^{3/2}\{ \frac{5}{3} F_{3/2}(\eta) - \eta F_{1/2}(\eta)\} . \tag{4.37}$$

Numerical values of $F_{1/2}$ and $F_{3/2}$ are listed in Table 4-1, from which the corresponding values of n, P, and s are evaluated. The mass density of the system is $\rho = mn$, where m is the neutron mass. For η exceeding 20, the following asymptotic expressions are sufficiently accurate: (η large)

$$F_{1/2}(\eta) \rightarrow \frac{2}{3}\eta^{1/2}\{ \eta + (\pi^2/8)\eta^{-1} + (7\pi^4/640)\eta^{-3} + \ldots\},$$

$$F_{3/2}(\eta) \rightarrow \eta^{1/2}\{ \frac{2}{5}\eta^2 + (\pi^2/4) - (7\pi^4/960)\eta^{-2} + \ldots\}.$$

The equations of state for a dense free neutron gas at constant temperatures, called isotherms, are plotted in Figure 4-1. The temperatures correspond to k_BT = 1, 3, 5, 7 and 10 MeV. It is apparent

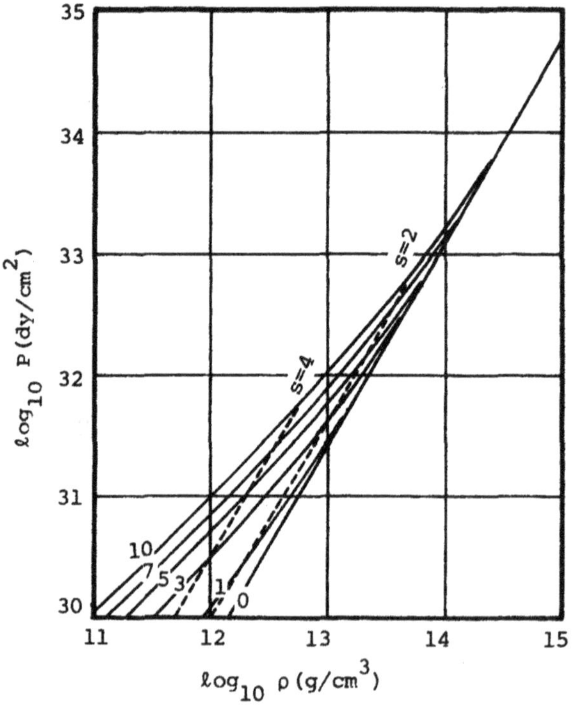

Figure 4-1. Equations of state for free neutron gas. Solid lines are isotherms at k_BT=1,3,5,7,10 MeV. Dotted lines are adiabats at s/k_B=2 and 4.

from these curves that temperature has negligible effects at high densities, for which the corresponding chemical potentials are much higher than k_BT. On the other hand, at low densities, the equation of state is greatly modified. In general, temperature effects tend to reduce the adiabatic index, defined in (2.30), of the equation of state making it "softer" than that at T=0.

The equation of state for an adiabatic process is called an adiabat. We shall be interested in adiabats for which the entropy per nucleon remains constant. They are given by:

$$s/k_B = S/k_B N = \frac{5}{3}\frac{F_{3/2}(\eta)}{F_{1/2}(\eta)} - \eta \ . \tag{4.38}$$

Curves of entropy per nucleon versus matter density at fixed temperatures are plotted in Figure 4-2, from which the adiabats may be extrapolated. These are shown as dotted curves in Figure 4-1. The adiabats are much "harder" than the isotherms as to be expected. We shall return to a more careful examination of these results later in Section 10.

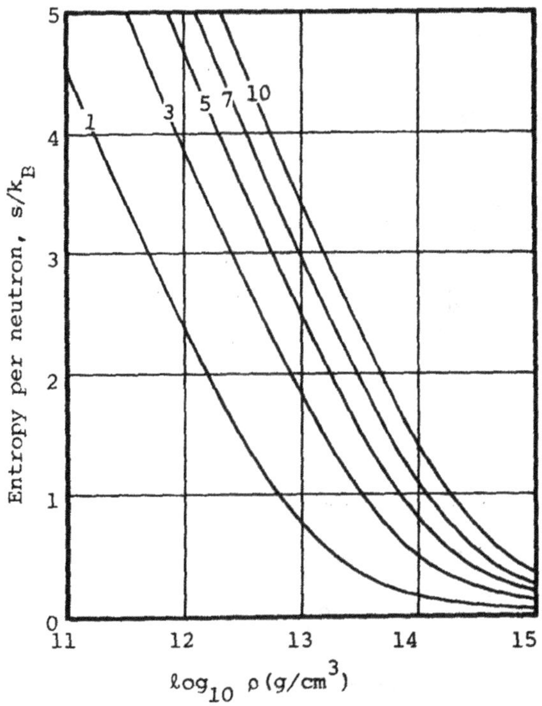

Figure 4-2. Entropy per neutron in a free neutron gas at temperatures given by k_BT=1,3,5,7,10 MeV.

So far we have discussed the solution of the many-body problem with interactions at zero temperature, and the case with no interaction at finite temperature. We shall now turn to the solution of the many-body problem with interactions and at finite temperature. In quantum statistics, the grand partition function is given by:

$$Z = \sum_n < \Psi_n | e^{\beta(\mu\hat{N} - \hat{H})} | \Psi_n > , \tag{4.39}$$

where $\hat{N}$ and $\hat{H}$ are the number operator and the Hamiltonian, respectively, and the summation over n refers to a summation over all orthogonal states of Hilbert spaces of any number of particles. An important variational principle for the partition function states that:

$$Z \geq Z_0 = \sum_n \exp\{\beta(\mu N_{nn} - H_{nn})\}, \tag{4.40}$$

where

$$N_{nn} = < \Psi_n | \hat{N} | \Psi_n > ,$$

and

$$H_{nn} = < \Psi_n | \hat{H} | \Psi_n > .$$

The best evaluation of Z for any set of basis functions Ψ_n corresponds to that which maximizes Z_0. Making use of the Hamiltonian (1.2) and for Ψ_n given by the Hartree-Fock trial wave functions, Z_0 may be expressed as:

$$Z_0 = \sum_{n_1=0}^{1} \sum_{n_2=0}^{1} \cdot \cdot \cdot \cdot \sum_{n_\infty=0}^{1} \exp\{-\beta(\sum_i (b_i-\mu)n_i + \frac{1}{2}\sum_{i,j} d_{ij}n_in_j)\} , \tag{4.41}$$

where b_i and d_{ij} are given by (1.13) and (1.22). Furthermore, single-particle wave functions appearing in d_{ij} belong to the same Hartree-Fock wave functions for a system of a definite number of particles N, since superselection rules will forbid the transition of one N state to a different N' state. In the Hartree-Fock approximation, the single-particle states are uncorrelated and it is permissible to express Z_0 in terms of the occupation probabilities $\bar{n}_i$, so that,

$$\ln Z_0 = -\beta \{ \sum_i (b_i-\mu)\bar{n}_i + \frac{1}{2} \sum_{i,j} d_{ij} \bar{n}_i \bar{n}_j \} . \tag{4.42}$$

To obtain the maximized $\ln Z_0$, the right-hand-side of (4.42) is first varied with respect to $\bar{n}_i$ keeping b_i and d_{ij} fixed. The condition for a maximum is found to be:

$$\bar{n}_i = \{ 1 + \exp \beta\left((b_i-\mu) + \sum_j d_{ij} \right)\}^{-1} , \qquad (4.43)$$

On the other hand, if the occupation probabilities $\bar{n}_i$ are kept fixed but variations are made in the single-particle wave functions as it is done in (1.10), we obtain the following thermal Hartree-Fock equations:

$$\{- \frac{\hbar^2}{2m} \nabla^2 + V(\vec{x}) + \sum_{j=1}^{N} \bar{n}_j \int d^3y\, \phi_j^*(\vec{y}) v(\vec{x}-\vec{y}) \phi_j(\vec{y}) \}\phi_i(\vec{x}) +$$

$$- \sum_{j=1}^{N} \delta(\sigma_i,\sigma_j) \bar{n}_j \int d^3y\, \phi_j^*(\vec{y}) v(\vec{x}-\vec{y}) \phi_i(\vec{y})\, \phi_j(\vec{x}) = \varepsilon_i^{HF} \phi_i(\vec{x}) . \qquad (4.44)$$

Applications of the thermal Hartree-Fock method will be carried out in Section 10.

The thermal Thomas-Fermi method for the atomic problem may be deduced by employing (4.43). The single-electron energies will be given by:

$$\varepsilon = \frac{p^2}{2m} - e\,\Phi ,$$

where Φ is the electrostatic potential satisfying the Gauss' law (3.7). The electron distribution will be approximated by $\bar{n}_p$ of (4.43), and the Gauss' law gives: (for $r \neq 0$)

$$\nabla^2 \Phi = (16\pi^2/h^3) e (2mk_BT)^{3/2}\, F_{1/2}(\beta e\Phi+\eta) . \qquad (4.45)$$

By a change of variable

$$r = cx ,$$

where

$$c = \left(32\pi^2 e^2 m (2mk_BT)^{1/2}\right)^{-1/2} ,$$

and

$$(\beta e\phi+\eta) = c\,\frac{u}{r} ,$$

(4.45) now reads:

$$\frac{d^2u}{dx^2} = x\, F_{1/2}\left(\frac{u}{x}\right) . \qquad (4.46)$$

The boundary conditions are:

(1) $x \to 0$: $u \to \beta\left(\frac{Ze^2}{c}\right)$,

(2) $x = r_s/c$: $\frac{du}{dx} = \frac{u}{x}$. (4.47)

Numerical solutions to this set of equations may be found in similar manner to the zero temperature case. It is discussed in article by Feynman, Metropolis and Teller (1949).

Table 4-1. Numerical values of $F_{1/2}(\eta)$ and $F_{3/2}(\eta)$ from McDougall and Stoner (1938), and Beer, Chase and Choquard (1955).

η	$F_{1/2}(\eta)$	$F_{3/2}(\eta)$
-4.0	.016128	.0242685
-3.0	.043366	.0656115
-2.0	.114588	.175800
-1.0	.290501	.460848
0.0	.678094	1.15280
1.0	1.39638	2.66168
2.0	2.50246	5.53725
3.0	3.97699	10.3537
4.0	5.77073	17.6277
5.0	7.83797	27.8024
6.0	10.1443	41.2610
7.0	12.6646	58.3422
8.0	15.3805	79.3526
9.0	18.2776	104.574
10.0	21.3445	134.270
11.0	24.5718	168.688
12.0	27.9518	208.062
13.0	31.4775	252.616
14.0	35.1430	302.564
15.0	38.9430	358.112
16.0	42.8730	419.458
17.0	46.9286	486.794
18.0	51.1061	560.305
19.0	55.4019	640.171
20.0	59.8128	726.568

References

Beer, A.C., Chase, M.N. and Choquard, P.F. (1950). Helv. Phys. Acta 28, 529.

Feynman, R.P., Metropolis, N. and Teller,E. (1949). Phys. Rev. 75, 1561.

McDougall, J. and Stoner, E.C. (1938). Phil. Trans. Roy. Soc, (London) A237, 350.

Bibliography

Huang, K. (1965). Statistical Mechanics, Wiley, New York.

Kittel,C. (1958). Elementary Statistical Physics, Wiley, New York.

Landau, L.D. and Lifshitz, E.M. (1958). Statistical Physics, Addison Wesley, Reading, Mass.

Thouless, D.J, (1961). The Quantum Mechanics of Many-Body Systems, Academic Press, New York.

CHAPTER 2

Regime II: $10^8 < \rho < 10^{14}$ g/cm^3. Subnuclear Matter

5. The Neutronization Process

Stellar matter in the density range of 10^8 to 10^{14} g/cm^3 is usually the result of extremely intense gravitational contraction and consequently subject to such high temperature that all nuclear exothermic reactions would have occurred in the process and all nuclides with atomic numbers below that of iron (Fe^{56}) would have burned themselves into a cinder of Fe^{56} nuclei. We use the formation of stellar matter composing of Fe^{56} nuclei to mark the termination of density Regime I. At higher densities new physics enters into the picture and it is naturnal to delineate a second density regime on the basis of the physics involved. This will be Regime II which is the main topic of Chapter 2. At densities above 10^8 g/cm^3 the constituent nuclei composing the matter transmute from those of Fe^{56} to more massive nuclides such as Ni, Ge, Zr, Sn, etc. The physical process is known as neutronization and is a direct consequence of the inverse β-decay reaction which proceeds at the rate of the weak interactions. The consequences of this process will be the subject of study in this chapter.

We have seen in the last chapter that degenerate electrons in matter become extremely energetic as soon as matter densities exceed 10^4g/cm^3. They become relativistic at approximately 10^6 g/cm^3. At 10^8 g/cm^3 their energy reaches that corresponding to the mass difference between a neutron and a proton which is given by Δmc^2:

$$m_n c^2 = 939.553 \text{ MeV}$$
$$m_p c^2 = 938.259 \text{ MeV}$$
$$\Delta mc^2 = 1.294 \text{ MeV} \quad (5.1)$$

Hence, protons become unstable via the weak interaction:

$$e + p \rightarrow n + \nu \quad (5.2)$$

where e, p, n, ν stand for electron, proton, neutron and neutrino, respectively.

For matter composes of electrons and nuclei, which we shall denote by (Z,A) where Z is the proton number and A the mass number, the nuclei will reduce their charge number by one unit while holding their mass number fixed under this reaction. However, the resulting nuclei may be metastable, and further transmutations will follow. For example, the following sequence of transmutations can be envisioned: $Fe^{56} \rightarrow Mn^{56} \rightarrow Cr^{56}$, simply because Cr being an even-even nuclide is more stable than Mn which is an odd-odd nuclide. Also, matter formed from Cr nuclei may not be the most favorable ground state at that particular density and if conditions were right Cr may further be transmuted into Ni to reach the true ground state. Detail mechanisms of these processes require precise knowledge of nuclear physics which we shall study with some care in this chapter. In general, nuclei become rich in neutrons as matter density increases, and thus the term neutronization. Also, new specis of nuclei more massive than Fe, some of which are unstable under terrestrial conditions, become the basic constituents of matter. We shall refer matter in the density range of 10^8 to 10^{14} g/cm^3 <u>subnuclear matter</u>.

Some insights into the problem and the method of attack may be acquired by studying the following hypothetical situation. Let us consider a simple system of electrons, protons and neutrons with all interactions turned off except for weak interaction (5.2). This is a problem of chemical equilibrium among these three constituents with the possibility of transmutations of protons into neutrons and vice versa. The pertinent entities of the problem in the ideal (relativistic) Fermi gas approximation are:

(1) number densities, $n_a = \frac{8\pi \, p_{Fa}^3}{3 \, h^3}$

(2) energy densities, $\varepsilon_a = \frac{8\pi}{h^3} \int_0^{p_{Fa}} p^2 dp \, (p^2c^2 + m_a^2c^4)^{\frac{1}{2}}$

(3) chemical potentials, $\mu_a = (p_{Fa}^2c^2 + m_a^2c^4)^{\frac{1}{2}}$ (5.3)

where subscript a denote e, p, n, and p_{Fa} are the Fermi momenta for the constituents. Due to charge neutrality, we have $n_e = n_p$, or $p_{Fe} = p_{Fp}$, and since n_p+n_n is conserved under the transmutation between p and n, it is convenient to introduce a conserved variable $n_B = n_p+n_n$, called the nucleon (or baryon) number density. We shall eliminate n_p from the expressions in favor of n_B and n_n. The total energy density of the system is given by:

$$\varepsilon = \varepsilon_n + \varepsilon_p + \varepsilon_e = \frac{8\pi}{h^3}\int_0^{p_{Fa}} p^2 dp\ (p^2c^2+m_a^2c^4)^{1/2} \tag{5.4}$$

Conditions for chemical equilibrium for a system at zero temperature may be obtained by minimizing ε,

$$\frac{\partial\varepsilon}{\partial n_n} = 0\ , \tag{5.5}$$

which yields:

$$\sum_a \left(\frac{\partial p_{Fa}}{\partial n_n}\right) \frac{8\pi}{h^3} p_{Fa}^2 (p_{Fa}^2c^2+m_a^2c^4)^{1/2} = 0 \tag{5.6}$$

Since

$$\frac{8\pi}{h^3} p_{Fa}^2 \left(\frac{\partial p_{Fa}}{\partial n_n}\right) = \pm\frac{1}{3}\ , \quad \text{with} \begin{array}{l} + \text{ for } a = n \\ - \text{ for } a = p, e \end{array} \tag{5.7}$$

(5.6) reduces to:

$$(p_{Fn}^2+m_n^2c^4)^{1/2} = (p_{Fp}^2+m_p^2c^4)^{1/2} + (p_{Fe}^2+m_e^2c^4)^{1/2} \tag{5.8}$$

which expresses the condition of chemical equilibrium in terms of the chemical potentials of the constituents:

$$\mu_n = \mu_p + \mu_e\ , \tag{5.9}$$

as would be derived by minimizing the Gibbs potential in (4.28). Since $p_{Fp}=p_{Fe}$, (5.8) is a relation between p_{Fn} and p_{Fe}. Once they are determined the relative number densities between neutrons and electrons are known.

The equation of state for such a system can be computed by writing

$$\rho = (m_p+m_e)n_e + m_n n_n = (m_p/Y_e)n_e\ , \tag{5.10}$$

and for the pressure we may take the extreme relativistic approximation of the degenerate electron pressure by replacing v by c in (2.19):

$$P = \frac{8\pi c}{3h^3}\int_0^{p_{Fe}} p^3 dp = \frac{2\pi c}{3h^3} p_{Fe}^4 = \hbar c\left(\frac{\pi^{2/3}}{12}\right)n_e^{4/3} \tag{5.11}$$

Had we started out with a pure system of protons and electrons before the onset of the weak interaction (5.2), an increase in matter density corresponds to an increase in proton number density as illustrated in Figure 5-1. Then, when matter density reaches $\rho \simeq 1.6 \times 10^7$ g/cm^3, the proton density will cease to increase as a finite neutron density emerges, which eventually overtakes the proton density. This

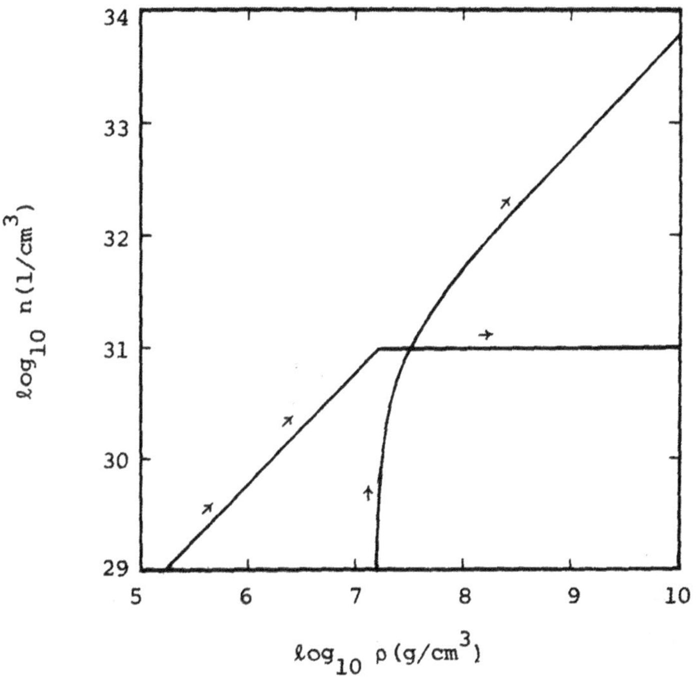

Figure 5-1. Proton and neutron number densities before and after the onset of weak interaction (5.2).

phenomenon has also profound effects on the equation of state of the system, which is shown in Figure 5-2. There we see that the pressure of the system remains a constant over two order of magnitudes of the matter densities.

The above study oversimplifies the real situation on several accounts. In reality, we are dealing with rather massive nuclei instead of protons and neutrons. Chemical equilibrium of the system will be determined by the mass differences of neighboring nuclides instead of the neutron-proton mass difference. Therefore, some reliable methods

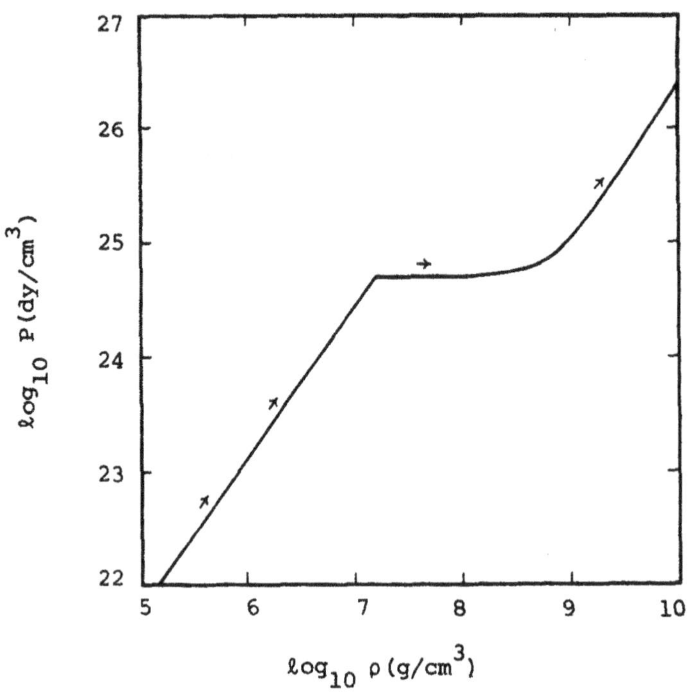

Figure 5-2. Equation of state of a system which starts out at low densities consisting entirely of protons and electrons.

in estimating the mass differences among nuclides, stable or unstable under terrestrial conditions, must be devised. To a large extend, this will be the main topics of discussion of this chapter.

Bibliogrpahy

Canuto, V. (1974). Ann. Rev. Astron. Astrophys. 12, 167.

Weinberg, S. (1972). Gravitation and Cosmology, Wiley, New York.

6. Nuclear Semi-empirical Mass Formula and Nuclear Mass Tables

Nuclear interactions are known to be very rich in details and are sufficiently strong in strength as to make perturbative calculations difficult. In spite of all the complexity associated with the interactions, stable nuclei exhibit patterns which can be understood in simple models. Nuclei have relatively well-defined sizes. The mass density of a nucleus has been found to be quite uniform over the entire volume, and is so similar among all stable nuclei that it suggests a common substance, call the nuclear matter, to describe all cases. The nuclei appear to keep on swelling as their mass increarses so as to maintain a common density like an imcompressible medium. This property of the nuclei resembles the saturation phenomenon in the sense that whenever the density is higher than the common density, it becomes unstable as in over-saturation and quickly returns to the common density. Hence, such a density is referred to as the saturation density or nuclear matter density. Approximating the nucleus as a sphere, the nuclear radius R can simply be expressed as:

$$R = r_0 \, A^{1/3} \, , \tag{6.1}$$

where r_0=1.12 fm (fermi, 1 fm=10^{-13} cm) and A is the mass number of the nucleus. The nuclear matter density ρ_0 can be found by relating it to r_0:

$$\rho_0 = (3/4\pi) m_p r_0^{-3} \simeq 2 \times 10^{14} \text{ g/cm}^3 \, . \tag{6.2}$$

There will be a surface to each nucleus. The mass density of the nucleus drops from 90% to 10% of ρ_0 over a surface thickness of $t_0 \simeq$ 2.4 fm, which is approximately the same for nuclei of all sizes. An approximate analytic form of the density distribution in a nucleus is given by:

$$\rho = \rho_0 \{ 1 + e^{(r-R)/D} \}^{-1}, \tag{6.3}$$

where r is the radial distance from the center of the nucleus and D is related to the surface thickness by t_o=2D(ℓn 9). This is illustrated in Figure 6-1.

A particular specis of nucleus, or nuclide, is labelled by its charge number Z and mass number A in the form (Z,A). The mass of a

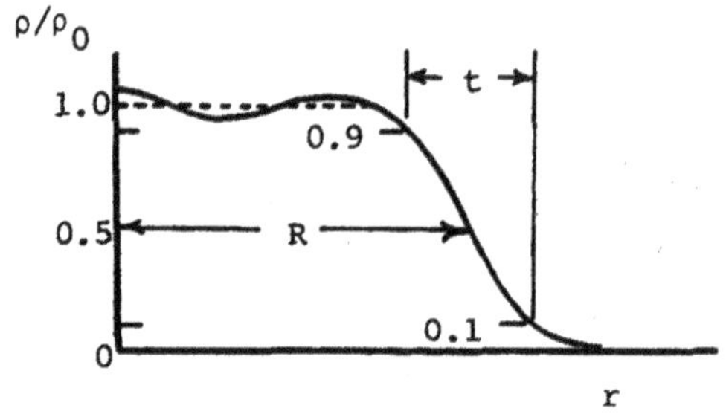

Figure 6-1. Approximate density distribution in a nucleus. The central portion of (6.3) is indicated by dotted lines.

nuclide M(Z,A) is determined by its total binding energy B(Z,A) as follows:

$$M(Z,A) = (m_p+m_e)Z + m_n(A-Z) - B(Z,A) . \qquad (6.4)$$

Empirically, we find that,

$$\frac{B(Z,A)}{A} \simeq 8.5 \text{ MeV} , \qquad (6.5)$$

which is true for all nuclides from A=12 and up. These are certainly very remarkable properties. The approximate proportionality of nuclear volume and mass number is commonly referred to as "saturation of nuclear density", and the approximate proportionality between total binding energy and mass number as "saturation of nuclear binding energies".

Nuclear Semi-empirical Mass Formula

A more accurate approximation of the nuclear binding energy for all stable nuclides can be expressed by the following semi-empirical mass formula:

$$B(Z,A) = a_V A - a_S A^{2/3} - a_C Z^2 A^{-1/3} - a_A \frac{(A-2Z)^2}{A} - \lambda a_P A^{-3/4} . \qquad (6.6)$$

The first term with coefficient a_V being proportional to A is called the volumetric term. It expresses a bulk property of the nuclear matter. The second term with coefficient a_S is called the surface energy term. The surface energy is proportional to the surface area, and with nuclear radius R proportional to $A^{1/3}$ as in (6.1), the surface area is proportional to $A^{2/3}$ as shown. The third term with coefficient a_C is called the Coulomb energy term. It accounts for the electrostatic energy of Z protons distributed uniformly over a spherical volume of radius R as derived for (3.4).

The fourth term with coefficient a_A is called the symmetry energy term. It is a nuclear effect expressing the fact that stable nuclides tend to have the same number of neutrons and protons, or A=2Z, and deviations from this symmetry will decrease the binding energy. It is quadratic in the asymmetry factor (A-2Z), since it must be positive and cannot be given by odd powers of the asymmetry factor. Also, it is treated as a bulk property so that it is also proportional to A. Together, they give a form as expressed in (6.6).

Furthermore, stable nuclides tend to have even numbers of protons and neutrons. For odd-A nuclides there is at most one stable isobar, while for even-A nuclides there may be two or more stable isobars. An isobar is a general term referring to nuclides with a given A. Even-even (both proton and neutron numbers are even) nuclides are more stable than even-odd or odd-odd nuclides. The differences in energy among these nuclides are included in the last term of the mass formula with:

$$\lambda = \begin{cases} +1 & \text{odd-odd} \\ 0 & \text{even-odd} \\ -1 & \text{even-even}\,, \end{cases} \qquad (6.7)$$

There term with coefficient a_P is called the pairing energy term. The $A^{-3/4}$ dependence is obtained empirically. The best-fit empirical formula for (6.6) has been found to be (Green 1954):

$$\begin{aligned} a_V &= 15.75 \text{ MeV} \\ a_S &= 17.8 \text{ MeV} \\ a_C &= 0.710 \text{ MeV} \\ a_A &= 23.7 \text{ MeV} \\ a_P &= 34 \text{ MeV}. \end{aligned} \qquad (6.8)$$

A plot of the binding energy expression (6.6) is shown in Figure 6-2, in which the relative magnitudes of some of the important terms in the semi-empirical mass formula are illustrated.

Let us now use the semi-empirical mass formula to obtain a qualitative determination of the neutronization process that might occur if the total energy of the system were maintained at the absolute minimum. Let us write the total energy density (including the mass of the nucleus) of the system as:

$$\varepsilon_T = \varepsilon_e + (n_B/A)\,\{M(Z,A)c^2 + \varepsilon_L\}\,, \qquad (6.9)$$

where ε_e is the degenerate electron energy density given by (2.21), n_B is the nucleon number density and (n_B/A) is the nuclei number density

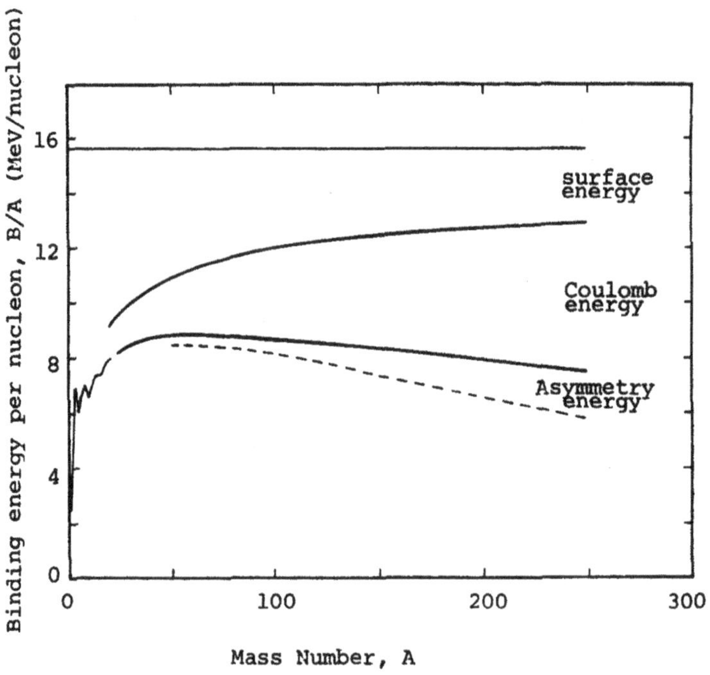

Figure 6-2. Relative contributions to the binding energy per nucleon as a function of the mass number A for the various terms in the semi-empirical mass formula.

in the system, M(Z,A) is given by (6.4), and ε_L is the electrostatic lattice energy of the nuclei per site with the assumption that the nuclei organize themselves into a lattice of dense packing. We shall ignore the lattice energy for the time being and concentrate on a qualitative study of the role of the nuclear mass M(Z,A) in the neutroization process. For the pairing energy term in the semi-empirical mass formula only the even-even nuclides which bind most strongly will be of interest to us and therefore we may take $\lambda=1$ in (6.6).

The total energy density of the system is a function of $\varepsilon_T=\varepsilon_T(Z,A,n_B)$. For any given n_B, ε_T would have to be minimized with respect to A and Z:

$$\text{(i)} \quad \left.\frac{\partial \varepsilon_T}{\partial z}\right|_{n_B} = 0 \,, \qquad \text{(ii)} \quad \left.\frac{\partial \varepsilon_T}{\partial A}\right|_{n_B} = 0 \,. \tag{6.10}$$

Condition (i) together with (2.18) for ε_e give:

$$\frac{\partial}{\partial Z}\{(n_B/A)M(Z,A) + (\pi^2\hbar^3 c)^{-1}\int_0^{p_{Fe}} p^2 dp(p^2+m_e^2c^2)^{1/2}$$

$$= (n_B/A)\frac{\partial M}{\partial Z} + (\pi^2\hbar^3 c)^{-1}p_{Fe}^2(p_{Fe}^2+m_e^2c^2)^{1/2}\frac{\partial p_{Fe}}{\partial Z} = 0 ,$$

where

$$p_{Fe} = \hbar(3\pi^2 Zn_B/A)^{1/3} , \tag{6.11}$$

and,

$$\frac{\partial p_{Fe}}{\partial Z} = \frac{1}{3}\frac{p_{Fe}}{Z} .$$

Condition (i) reduces to:

$$(n_B/A)\frac{\partial M}{\partial Z} + (n_e/Zc)(p_{Fe}^2+m_e^2c^2)^{1/2} = 0 ,$$

or simply,

$$\frac{\partial M}{\partial Z} + (p_{Fe}^2/c^2 + m_e^2)^{1/2} = 0 . \tag{6.12}$$

Similarly, condition (ii) reduces to:

$$A\frac{\partial M}{\partial A} - M - Z(p_{Fe}^2/c^2 + m_e^2)^{1/2} = 0 . \tag{6.13}$$

By eliminating p_{Fe} from (6.12) and (6.13), they combine into:

$$Z\frac{\partial M}{\partial Z} + A\frac{\partial M}{\partial A} = M . \tag{6.14}$$

Straight forward partial differentiations of (6.6) provide the relation between Z and A which yields the condition of energy minimization at fixed n_B. It is given by:

$$a_S A^{2/3} - 2a_C Z^2 A^{-1/3} + (4/3)a_P A^{-3/4} = 0 ,$$

or,

$$Z = \{(a_S/a_C)A/2 + (3a_P/8a_C)A^{-5/12}\}^{1/2}$$

$$= 3.54\,A^{1/2}\{1 + 1.43A^{-17/12}\}^{1/2} \simeq 3.54\,A^{1/2} , \tag{6.15}$$

using values listed in (6.8).

Once Z as a function of A is known, any of the two relations (6.12) and (6.13) would give A as a function of n_B. One finds that A is a monotonic increasing function of n_B, which together with (6.15) shows that Z is also a monotonic increasing function of n_B. Hence, we deduce from the semi-empirical mass formula the following qualitative features of the neutronization process; as matter density increases constituent nuclei are becoming more and more massive and at the same time becoming more and more neutron rich. These features were first derived by Bethe, Borner, and Sato (1970). They predicted the occurrence of very massive nuclei with $Z=51$ and $A=211$ in dense matter.

Let us mention here that there are also several other versions of the semi-empirical mass formula in addition to (6.6). A partial list of references to some of the more recent proposals includes: (Seegar 1972), (Myers and Swiatecki 1974), (Myers 1976), (H. von Groote, et al. 1976). The qualitative behavior between Z and A predicted by these versions of the mass formula for the energy minimization problem is similar to (6.15), at least for A not too large. At very large A, $A > 100$, Buchler and Barkat (1971) showed that the use of the mass formula of Seegar (1968), which included a surface-symmetry term, gave a dependence of Z on A which exhibited an upper limit in Z at some value of A, and Z dropped in value at further increase of A. This feature, they claimed, is more realistic than the one predicted by (6.15) when compared to the more elaborate Hartree-Fock calculations. In any case, as long as the semi-empirical mass formula is applied to nuclides very different from the stable nuclides, the results will be sensitive to the functional form of the mass formula. In order to achieve some confidence in the results it is necessary to move on to other methods based on better theoretical models. A more up-to-date discussion of the status in the use of the semi-empirical mass formula for the present purpose is given by Lattimer (1981).

Nuclear Mass Tables

When we are dealing with nuclides which are not too different from the stable nuclides, there exists rather complete tables of nuclear masses which are based on actual experimental measurements or extrapolations from measurements on similar nuclides. With nuclear masses given by tables we can obtain a more accurate determination of the neutronization process than it can be done with the semi-empirical mass formula. This approach is applicable to matter with densities in the range between 10^8 to 4×10^{11} gm/cm^3.

The total energy density ε_T to be minimized is given by (6.9). The lattice energy ε_L has been computed by Coldwell-Horsfall and Maradudin (1960) for the following dense packing lattices:

$$\varepsilon_L = -\frac{(Ze)^2}{r_s} \times \begin{cases} 0.89593 & \text{BCC (Body Centered Cubic)} \\ 0.89588 & \text{FCC (Face Centered Cubic)} \\ 0.88006 & \text{HCP (Hexagonal Close Pack)} \end{cases} \qquad (6.16)$$

where $(4\pi/3)r_s^3 = (n_B/A)^{-1}$. ε_L is computed by placing a nucleus of charge Z at each of the lattice points. They will be surrounded by a uniform background of negative charges. It is apparent that the BCC lattice provides the lowest energy, and let us therefore assume that it is indeed the configuration taken up by the nuclei.

For a fixed nucleon number density n_B, ε_T is a function of A and Z:

$$\varepsilon_T = (n_B/A)\,\{M(Z,A)c^2 - 1.4442(Ze)^2(n_B/A)^{1/3}\} + \varepsilon_e \,. \qquad (6.17)$$

For matter densities exceeding $\rho \simeq 10^6$ g/cm^3, extreme relativistic approximation for the electron energy would be appropriate and ε_e may be written as:

$$\varepsilon_e \simeq (\hbar c)\,(3/4)\,(3\pi^2 Z n_B/A)^{1/3} \,. \qquad (6.18)$$

The nucleon number density at a matter density of $\rho = 10^6$ g/cm^3 is approximately $n_B = 10^{-9}$ fm^{-3}. Expressing n_B in units of $n_{BO} = 10^{-9}$ fm^{-3}, the <u>average energy per nucleon</u>, (ε_T/n_B), in the system is given by (in units of MeV):

$$\varepsilon_T/n_B = M(Z,A)c^2/A + (0.4578Z^{4/3} - \frac{Z^2}{480.74})\,(n_B/n_{BO})^{1/3}\,A^{-4/3}. \qquad (6.19)$$

Tables of nuclear masses are given by Garvey et al. (1969). The average energy per nucleon in a nucleus is approximately 930 MeV. By writing $M(Z,A)c^2/A = (930+\Delta)$ MeV, the values of Δ according to Garvey et al. for even-even nuclides are listed in Table 6-1 (for Fe and above). These are candidate nuclides which would achieve energy minimization in a neutronization process.

One computes for a nuclide the quantity (ε_T/n_B) according to (6.19) as a function of n_B, and plots it versus $(1/n_B)$ as in Figure 6-3. When several such curves belonging to different nuclides in the same range of n_B are plotted and compared, there is no difficulty in picking out the curve which gives the minimum energy at that density. The nuclide to which the curve belongs is the one which satisfies the minimum condition and constitutes the stable ground state of the system at that density.

Table 6-1. Nuclear mass table for some even-even nuclides. Values given are $\Delta = M(Z,A)c^2/A - 930$ in MeV.

Nuclide Z / A-Z	Fe 26	Ni 28	Zn 30	Ge 32	Se 34	Kr 36	Sr 38	Zr 40	Mo 42	Ru 44
30	.2079	.2511								
32	.2227	.2158								
34	.2669	.2150								
36	.3327	.2396	.2492							
38	.4195	.2921	.2609							
40	.5287	.3629	.2944							
42		.4449	.3456	.3023						
44		.5381	.4044	.3278						
46		.6307	.4662	.3700	.3193					
48	1.0119	.7231	.5404	.4173	.3440	.3099	.3089			
50	1.1314	.8269	.6186	.4740	.3781	.3206	.2933			
52	1.0943	.9839	.7543	.5868	.4670	.3829	.3377			
54			.8928	.7025	.5568	.4554	.3897			
56			1.0304	.8153	.6532	.5319	.4495			
58				.9385	.7572	.6199	.5180			
60				1.0631	.8667	.7104	.5932			
62				1.1913	.9768	.8061	.6762			
64				1.3183	1.0902	.9074	.7579			
66				1.4504	1.2109	1.0092	.8467	.7209		
68				1.5874	1.3297	1.1157	.9389	.7991		
70					1.4510	1.2233	1.0332	.8832		
72					1.5725	1.3322	1.1325	.9697		
74					1.6930	1.4438	1.2320	1.0556		
76					1.8153	1.5548	1.3299	1.1416		
78					1.9349	1.6620	1.4261	1.2273	1.0629	
80					2.0504	1.7671	1.5214	1.3116	1.1360	1.1972
82					2.1649	1.8725	1.6162	1.3958	1.2099	1.0462
84						2.0061	1.7395	1.5087	1.3131	1.1417
86						2.1371	1.8606	1.6208	1.4178	1.2390
88						2.2671	1.9822	1.7355	1.5254	1.3393

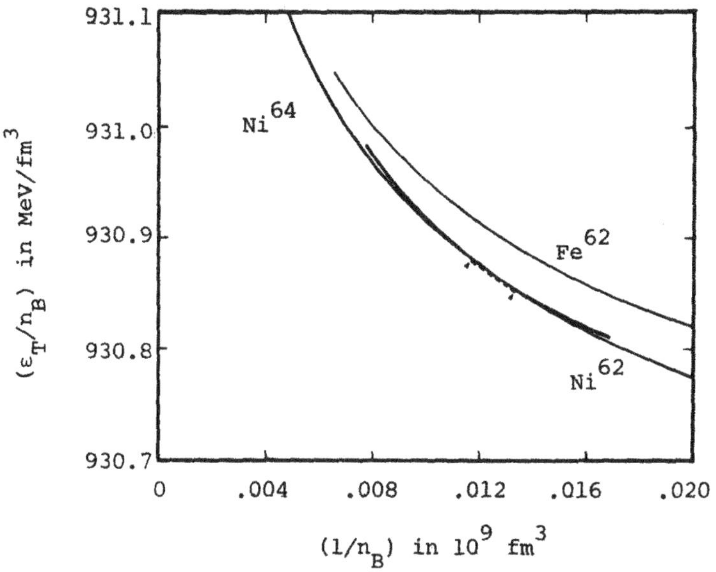

Figure 6-3. Energy curves according to (6.19) for nuclides Fe^{62}, Ni^{62}, and Ni^{64}. The short-dashed line is a tangent to the Ni^{62} and Ni^{64} curves. The triangular pointers indicate the points of tangency.

As density increases a second nuclide would take over in forming the stable ground state. This happens when two curves lying on the envelope of all the curves cross each other. A first order phase transition is taking place in the neighborhood of the point of crossing. This is shown by the crossing of the Ni^{62} curve by the Ni^{64} curve in Figure 6-3. The exact density at which phase transition occurs and the density at which it terminates may be found from a tangent construction method. This is also shown in Figure 6-3. By this method one simply draws a straight line tangent to the convex curves belonging to these nuclides. The values of n_B at the points of tangency correspond to the onset and termination of the phase transition. In a first order phase transition, the pressure of the system remains constant as the density varies. Now, all points on the tangent in the above construction have the same pressure, since pressure is expressible as the slope of a straight line on such a plot. This can be seen as follows:

$$P = -\left.\frac{\partial E}{\partial \Omega}\right|_N = -\left.\frac{\partial(\varepsilon_T \Omega)}{\partial \Omega}\right|_N = -\frac{\partial(\varepsilon_T/n_B)}{\partial(1/n_B)} , \qquad (6.20)$$

where the last equality comes from the fact that $\Omega \propto n_B^{-1}$ at fixed total number of nucleons N.

Once the nuclide (Z,A) is selected for a specific range of densities, the equation of state for that range can be evaluated. The matter density is given by:

$$\rho = M(Z,A)\, n_B / A , \qquad (6.21)$$

and the pressure computed from (2.22) with t given by:

$$t = \lambda_e \left(\frac{3\, Z\, n_B}{8\pi\, A} \right)^{1/3} , \qquad (6/22)$$

where $\lambda_e = 2.426\text{x}10^{-10}$ cm is the electron Compton wavelength.

The equation of state for matter at zero temperature and in complete nuclear equilibrium taking into account lattice effects and making use nuclear masses in the manner described in this section has been derived by Baym, Pethick and Sutherland (1971) for matter densities up to $\rho = 4.3 \times 10^{11}$ g/cm^3. They use the nuclear mass tables of Myers and Swiatecki (1966) given in an unpublished report. There are slight differences in the nuclear energies used by them and those listed in Table 6-1, but these are not large enough to affect significantly the equation of state. We list in Table 6-2 the sequence of nuclides which they found to form the ground state of subnuclear matter, and the maximum densities ρ_{max} for which they occur. In Table 6-2 the nuclear mass employed by these authors are also listed. They are to be compared with those listed in Table 6-1 for the same (Z,A). The equation of state given by these authors are shown in Figure 6-4. Plotted on the same graph for comparison is the equation of state of a non-interacting (free) degenerate gas of electrons, protons and neutrons with $Y_e = Z/A = 0.5$.

There are, however, limitations to this type of approach. Some of the nuclei employed in this scheme have Z/A ratios so different from those of the stable nuclei (typically, $Z/A \simeq 1/2$), that it is not certain how accurate their extrapolated masses given in the mass table are, and caution should be excercised. Also, at densities higher than $\rho \simeq 4.3\text{x}10^{11}$ g/cm^3 the phenomenon called "neutron drip" occurs, which describes a situation where there is negligible change in the binding energy of a nucleus when an extra neutron is added to or taken away from it. Neutrons are no longer bound to nuclei and appear to be "dripping" out of the nuclei. These nuclei coexist with a sea of neutrons. In order to understand this and other related phenomena, the use of the nuclear mass tables will not be adequate, and it is necessary to develop theoretical metnods which can on the one hand reproduce the results of the nuclear mass tables and on the other be capable to handle the neutron drip problem. These methods will be discussed in the following sections.

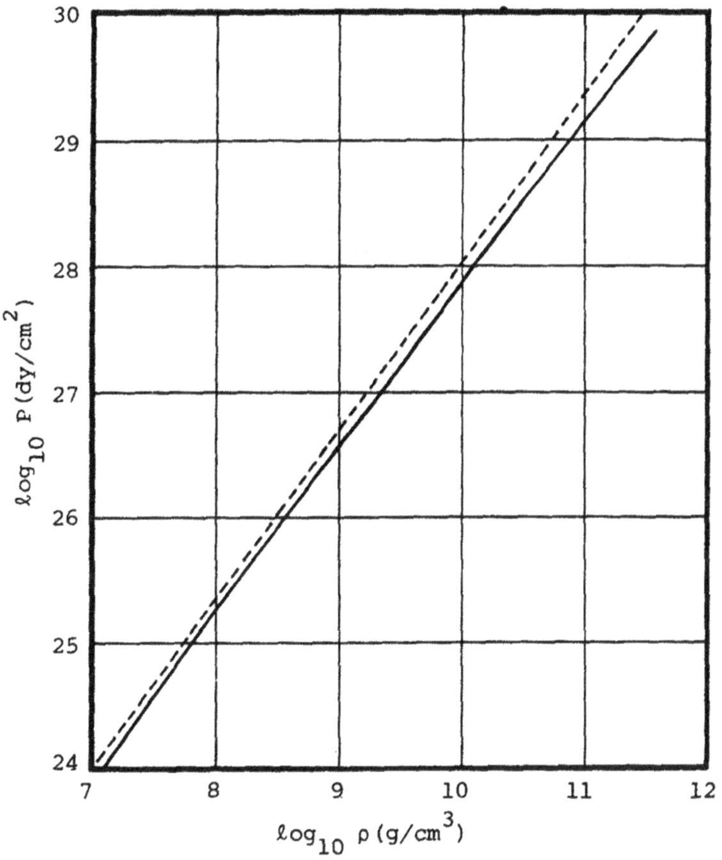

Figure 6-4. The equation of state of subnuclear matter at zero temperature given by Baym, Pethich and Sutherland (1971). The equation of state of a non-interacting degenerate gas of electrons, protons and neutrons with Y_e=0.5 is plotted in dotted lines.

Table 6-2. Sequence of nuclei forming the ground state of matter and the maximum densities to which they occur. The nuclear masses of these nuclides are given by $\Delta = M(Z,A)c^2/A - 930$ in MeV.

Nuclide	Z	(A-Z)	Z/A	Δ (MeV)	ρ_{max} (g/cm^3)
Fe^{56}	26	30	.4643	.1616	8.1E6
Ni^{62}	28	34	.4516	.1738	2.7E8
Ni^{64}	28	36	.4375	.2091	1.2E9
Se^{84}	34	50	.4048	.3494	8.2E9
Ge^{82}	32	50	.3902	.4515	2.2E10
Zn^{84}	30	54	0.3750	.6232	4.8E10
Ni^{78}	28	50	.3590	.8011	1.6E11
Fe^{76}	26	50	.3421	1.1135	1.8E11
Mo^{124}	42	82	.3387	1.2569	1.9E11
Zr^{122}	40	82	.3279	1.4581	2.7E11
Sr^{120}	38	82	.3166	1.6909	3.7E11
Kr^{118}	36	82	.3051	1.9579	4.3E11

References

Baym G., Pethick, C. and Sutherland, P. (1971). Astrophys. J. 170, 299.

Bethe, H.A., Borner, G. and Sato, K. (1970). Astron. Astrophy. 7, 279.

Buchler, J.-R. and Barkat, Z. (1971). Astrophys. Lett. 7, 167.

Coldwell-Horsfall, R.A. and Maradudin, A.A. (1960). J. Math. Phys. 1, 395.

Garvey, G.T., Gerace, W.J., Jaffe, R.L., Talmi, I. and Kelson, I. (1969). Rev. Mod. Phys. 41, sl.

Green, A.E.S. (1954). Phys. Rev. 95, 1006.

Lattimer, J.M. (1981). Ann. Rev. Nucl. Part. Sci. 31, 337.

Myers, W.D. and Swiatecki, W.J. (1974). Ann. Phys. 84, 211.

Myers, W.D. (1976). In Atomic Data and Nuclear Data Tables, (Ed. K. Way), Vol. 17, p.411. Academic Press, New York.

Seegar, P.A. (1968). In Amer. Inst. Phys. Handbook, (E. D.E. Gray), 3rd edition. McGraw-Hill, New York.

Seegar, P.A. (1972). Ibid.

Von Groote, H., El Eid, M.F. Takahashi, K. (1976). In Atomic Data and Nuclear Data Tables, (Ed. K. Way), Vol. 17, p.418. Academic Press, New York.

7. Nuclear Two-Body Interactions

Past experiences in dealing with the electromagnetic and gravitational interactions show that once the basic interactions between two elementary objects are known, the properties of a large conglomeration of such objects can also be predicted. The purpose of a many-body theory is to evaluate such properties starting with basic interactions. The search for the basic two-nucleon interactions had been a major preoccupation of the nuclear physicists ever since experimental nuclear physics became feasible. A great deal of the pertinent details about nuclear interactions has been revealed through scattering experiments and low energy nuclear studies. A first step in expressing low energy nuclear interactions is by constructing two-nucleon potentials. Once such nuclear potentials are available a many-body Hamiltonian in the form of (1.2) may be written down and a solution to the many-body problem may be attempted. Nuclear interactions can also be described in other ways appropriate to specific schemes of investigation, such as in terms of a scattering matrix, or a density matrix, or meson parameters. In this section, we shall limit our discussions to a description of the nuclear interactions through a parametrization of the nuclear potentials.

As a start, the nuclear forces may be assumed to be conservative and expressible as the gradients of potentials, which then should depend only on the coordinates of the nucleons and not on their momenta. We shall first look at such non-velocity dependent potentials, or static potentials. It is clear that the two-body potential functions cannot be dependent on the individual nucleon coordinates but only on their relative coordinates, denoted by $\vec{r}$.

Nuclear interactions are known to be short-ranged, and as suggested by the meson theory the nuclear potential $v(\vec{r})$ should be of the Yukawa type:

$$v(\vec{r}) \sim \frac{e^{-r/R}}{r} , \qquad (7.1)$$

where $r = |\vec{r}|$ is the internucleon separation, and R denotes the range of the potential. The range R should be of the order of the pion Compton wavelength, $\hbar/m_\pi c$. For computational convenience, other less singular short-range potentials, such as the square-well potential or Gaussian potential in the form $v \sim \exp(-r^2/R^2)$, are also commonly employed.

Allowance is made for the nuclear potentials to be dependent on the spin orientations of the nucleons. Since the assignment of the axis of quantization of the spin components is arbitrary, a pair of spin-up protons should not interact differently from a pair of spin-down protons. A spin dependent nuclear potential should possess this and other geometric properties. Fortunately such fine details are completely taken care of if the spin wave functions obey angular momentum rules and the spin-dependent potentials satisfy rotational invariance.

Let us denote the nucleon spin wave functions by χ_σ and represent the two components of χ_σ by two-valued column matrices:

<u>spin-up</u>: $\chi_\uparrow = \begin{pmatrix} 1 \\ 0 \end{pmatrix}$, <u>spin-down</u>: $\chi_\downarrow = \begin{pmatrix} 0 \\ 1 \end{pmatrix}$ (7.2)

The spin operators $\vec{s}$ are then represented by the Pauli matrices $\vec{\sigma}$:

$$\vec{s} = \frac{\hbar}{2}\vec{\sigma} \qquad (7.3)$$

where

$$\sigma_x = \begin{pmatrix} 0 & 1 \\ 1 & 0 \end{pmatrix}, \quad \sigma_y = \begin{pmatrix} 0 & -i \\ i & 0 \end{pmatrix}, \quad \sigma_z = \begin{pmatrix} 1 & 0 \\ 0 & -1 \end{pmatrix}. \qquad (7.4)$$

χ_σ satisfy the angular momentum rules as can be verified by matrix multiplications:

$$\vec{s}^{\,2}\chi_\sigma = s(s+1)\hbar^2\chi_\sigma , \qquad (s=\tfrac{1}{2})$$
$$s_z\chi_\sigma = \pm\tfrac{1}{2}\hbar\chi_\sigma . \qquad (7.5)$$

For a two-nucleon system, the single-nucleon spin wave functions may be combined symmetrically into the spin triplet states, or antisymmetrically into a spin singlet state. The triplet states are:

$$\chi_1^1 = \chi_\uparrow(1)\,\chi_\uparrow(2) ,$$
$$\chi_1^0 = \frac{1}{\sqrt{2}}\{\chi_\uparrow(1)\,\chi_\downarrow(2) + \chi_\downarrow(1)\,\chi_\uparrow(2)\} ,$$
$$\chi_1^{-1} = \chi_\downarrow(1)\,\chi_\downarrow(2) , \qquad (7.6)$$

and the spin singlet state:

$$\chi_0 = \frac{1}{\sqrt{2}}\{\chi_\uparrow(1)\,\chi_\downarrow(2) - \chi_\downarrow(1)\,\chi_\uparrow(2)\} . \qquad (7.7)$$

The total spin operator of the two-nucleon wave function is given by $\vec{S} = \vec{s}_1 + \vec{s}_2$, each component of which acts as follows:

$$\begin{aligned} S_z \chi_1^1 &= (s_{1z} + s_{2z}) \chi_\uparrow(1) \chi_\uparrow(2) \\ &= \{s_{1z} \chi_\uparrow(1)\}\chi_\uparrow(2) + \chi_\uparrow(1) \{ s_{2z} \chi_\uparrow(2)\} \\ &= \hbar \chi_\uparrow(1) \chi_\uparrow(2) \quad . \end{aligned}$$

which is a compact way of writing a direct product of matrices. In like manner, one can show in general:

$$\begin{aligned} S_z \chi_1^m &= m\hbar \chi_1^m \quad , \quad (m = 1, 0, -1) \\ S_z \chi_0 &= 0 \quad . \end{aligned} \tag{7.8}$$

For the total spin operator, its eigenvalues are evaluated as follows:

$$\begin{aligned} \vec{S}^2 \chi_1^1 &= (\vec{s}_1+\vec{s}_2)\cdot(\vec{s}_1+\vec{s}_2) \chi_1^1 = (s_1^2+s_2^2+2\vec{s}_1\cdot\vec{s}_2) \chi_1^1 = \\ &= \{\vec{s}_1^2\chi_\uparrow(1)\} \chi_\uparrow(2) + \chi_\uparrow(1)\{\vec{s}_2^2\chi_\uparrow(2)\} + \frac{\hbar^2}{2}\{\sigma_{1z}\chi_\uparrow(1)\}\{\sigma_{2z}\chi_\uparrow(2)\} + \\ &\quad + \frac{\hbar^2}{2}\{ \sigma_{1x}\chi_\uparrow(1)\}\{\sigma_{2x}\chi_\uparrow(2)\} + \frac{\hbar^2}{2}\{ \sigma_{1y}\chi_\uparrow(1)\}\{\sigma_{2y}\chi_\uparrow(2)\} = \\ &= \left(\frac{3}{4} + \frac{3}{4} + \frac{1}{2}\right) \hbar^2 \chi_\uparrow(1) \chi_\uparrow(2) + \left(+ \frac{1}{2} - \frac{1}{2}\right) \hbar^2 \chi_\downarrow(1) \chi_\downarrow(2) = \\ &= 2 \hbar^2 \chi_\uparrow(1) \chi_\uparrow(2) \quad . \end{aligned}$$

Thus, all three components of the triplet states satisfy the angular momentum rules for spin-1 states: (S=1)

$$\vec{S}^2 \chi_1^m = S(S+1) \hbar^2 \chi_1^m \quad , \quad (m = 1, 0, -1) \tag{7.9}$$

and the singlet state for spin-0 state:

$$S^2 \chi_0 = 0 \quad . \tag{7.9'}$$

The spin-dependence of the potentials will be based entirely on whether the two nucleons are in the triplet states or the singlet state, and not on the component index m, or in other words, based on whether the spin states are in symmetric or antisymmetric combinations. Hence, the spin-dependence of the potentials may also be accomplished by a spin exchange operator P_σ, whose action on a pair of spin wave functions (or linear combinations of them) is:

$$P_\sigma \, \chi_\sigma(1)\chi_{\sigma'}(2) \;=\; \chi_{\sigma'}(1)\chi_\sigma(2) \quad . \tag{7.10}$$

It follows immediately that:

$$P_\sigma \, \chi_1^m \;=\; \chi_1^m \; , \qquad (m = 1, 0, -1)$$

$$P_\sigma \, \chi_0 \;=\; -\chi_0 \quad .$$

It can also be verified that P_σ has the following realization:

$$P_\sigma \;=\; \frac{1}{2} \left(1 + \vec{\sigma}_1 \cdot \vec{\sigma}_2 \right) \tag{7.11}$$

Hence, spin-dependent potentials may be written as:

$$v \;=\; v_1(r) + v_2(r) \, P_\sigma \quad , \tag{7.12}$$

or, equivalently,

$$v \;=\; v_C(r) + \vec{\sigma}_1 \cdot \vec{\sigma}_2 \, v_S(r) \quad . \tag{7.13}$$

Experimental results indicate that, in so far as the Coulomb interaction may be neglected, a pair of protons interact in the same way as a pair of neutrons. The situation is very similar to the spin-dependent interactions, where two spin-up states and two spin-down states give the same interaction. It is therefore tempting to treat the proton and neutron as two states equivalent to the spin-up and spin-down states of a single spin vector obeying angular momentum rules. Such an approach is called the isospin (or isotopic spin) formalism. Nuclear interactions obeying the isospin formalism are called <u>charge independent</u>.

In the isospin formalism, the proton and neutron form the two components of an isospin-½ vector obeying angular momentum rules. The isospin wave functions are denoted by:

<u>proton</u>: $\zeta_p = \begin{pmatrix} 1 \\ 0 \end{pmatrix}$, <u>neutron</u>: $\zeta_n = \begin{pmatrix} 0 \\ 1 \end{pmatrix}$, (7.14)

and the isospin operators $\vec{t}$ by:

$$\vec{t} \;=\; \frac{1}{2} \, \vec{\tau} \quad , \tag{7.15}$$

where $\vec{\tau}$ are again the Pauli matrices acting in the isospin space. Proton and neutron are then the two states of a single entity, the nucleon. A two-nucleon system may again be combined symmetrically into three isospin triplet states, and antisymmetrically into an isospin singlet state in the same manner as it is done for the spin wave functions.

We shall assume that nuclear potentials are charge independent. The difference in interactions between the isospin triplet states and the singlet state may be introduced by an isospin exchange operator P_τ which acts on the isospin wave functions as follows:

$$P_\tau \; \zeta_q(1) \; \zeta_{q'}(2) \;=\; \zeta_{q'}(1) \; \zeta_q(2) \; . \tag{7.16}$$

P_τ may be written as:

$$P_\tau \;=\; \frac{1}{2} \left(1 + \vec{\tau}_1 \cdot \vec{\tau}_2 \right) . \tag{7.17}$$

Two-nucleon potentials which are spin and isospin dependent may be expressed as follows:

$$v \;=\; v_1 + v_2 \, P_\sigma + \left(v_3 + v_4 \, P_\sigma \right) P_\tau \; , \tag{7.18a}$$

or, equivalently,

$$v \;=\; c_C + v_S (\vec{\sigma}_1 \cdot \vec{\sigma}_2) + \{ v'_C + v'_S (\vec{\sigma}_1 \cdot \vec{\sigma}_2) \} \; (\vec{\tau}_1 \cdot \vec{\tau}_2) \; . \tag{7.18b}$$

The above potentials being functions of the internucleon separation r will give rise to nuclear forces which are central, i.e., directed along the relative coordinate $\vec{r}$. For a deuteron, which is a simple two-nucleon state, pure central nuclear forces would imply that its ground state should be in a zero orbital angular momentum state, or S state, since there is no cause to prevent it from occupying such a state. However, deuteron is found to possess a quadrupole moment implying it is not distributed spherically symmetrically as required by the S state configuration. The inevitable conclusion from this fact is that nuclear forces cannot be purely central and they must contain admixtures of non-central parts. The only non-central force which is velocity-independent is derivable from the following tensor force potential containing the tensor operator S_{12} give by: ($\hat{r} = \vec{r}/r$)

$$S_{12} \;=\; 3 \; (\vec{\sigma}_1 \cdot \hat{r}) (\vec{\sigma}_2 \cdot \hat{r}) \; - \; (\vec{\sigma}_1 \cdot \vec{\sigma}_2) \; . \tag{7.19}$$

The inclusion of a tensor force potential in the form of $S_{12} v_T(r)$ would then account for the qradrupole moment of the deuteron. The tensor operator has the following properties that it annihilates the two-nculeon spin singlet state:

$$S_{12} \, \chi_0 \;=\; 0 \; . \tag{7.20}$$

This can easily be verified by taking r to be z-direction and hence $S_{12} = 3\sigma_{1z}\sigma_{2z} - (\sigma_{1z}\sigma_{2z}+\sigma_{1x}\sigma_{2x}+\sigma_{1y}\sigma_{2y})$. Such an operator acting on the singlet χ_0 state yields a vanishing eigenvalue. The same result is obtained if $\hat{r}$ were taken to be either the x- or y-direction. Thus, only the triplet states contribute to the tensor force.

Nuclear potentials satisfying charge independence and with tensor interaction may be expressed as:

$$v = v_1 + v_2P_\sigma + v_5S_{12} + \{v_3 + v_4P_\sigma + v_6S_{12}\} P_\tau , \tag{7.21a}$$

or,

$$v = v_C + v_S(\vec{\sigma}_1\cdot\vec{\sigma}_2) + v_TS_{12} + \{v'_C + v'_S(\vec{\sigma}_1\cdot\vec{\sigma}_2) + v'_TS_{12}\} (\vec{\tau}_1\cdot\vec{\tau}_2) . \tag{7.21b}$$

For a two-nucleon state the interchange of the nucleon spatial coordinates, $\vec{x}_i\leftrightarrow\vec{x}_j$, is performed by an operator called the Majorana operator P_M in nuclear physics. Since all nucleons are treated as identical particles in the charge independent scheme, the two-nucleon state is therefore antisymmetric under the combined interchanges of the nucleon spins, isospins, and spatial coordinates. Hence, the operation of the Majorana operator on the two-nucleon states is equivalent to:

$$P_M = - P_\sigma P_\tau . \tag{7.22}$$

The potentials (7.18a,b) may also be parametrized in terms of P_M. Under the interchange of the nucleon spatial coordinates the relative coordinates change from $\vec{r}$ to $-\vec{r}$ which is a parity operation.

Static nuclear potentials in the form of (7.21a,b) have been proposed. The more successful ones include the Gammel-Thaler potentials (Gammel, Thaler 1957; see also Brueckner, Gammel, Thaler 1958). Subsequently, more elaborate potentials including velocity-dependent terms have been proposed. We shall study two-nucleon potentials further in Chapter 3. In the remainder of this section, we shall illustrate the application of the static potentials in a many-body calculation for nuclear matter.

Nuclear matter represents the ground state of a nucleus experiencing no surface or Coulomb effects. It may be considered to be a uniform degenerate Fermi system of infinite extend. It contains equal numbers of protons and neutrons, and each quantum state has a degeneracy factor of four, i.e., proton and neutron of spin-up and spin-down states. The saturation properties exhibited by stable nuclei make the study of nuclear matter meaningful. All nuclei no matter how large or small may be considered to consist of a body of nuclear matter perturbed somewhat by surface and Coulomb effects. Nuclear matter has been estimated to be at a saturation density of $\rho_0 = 2.8 \times 10^{14}$ g/cm^3, or a nucleon number density of $n_0 = 0.170$ nucleons/fm^3, and the average binding energy per nucleon is $\varepsilon_B = 15.68$ MeV (Bethe 1971).

We shall make an evaluation of the many-body equation (1.1) with the Hartree-Fock wave functions (1.17). The Hamiltonian H will be given by:

$$H = \sum_i -\frac{\hbar^2}{2m}\nabla_i^2 + \frac{1}{2}\sum_{i,j} v(\vec{x}_i-\vec{x}_j) \, , \tag{7.23}$$

where $v(\vec{x}_i-\vec{x}_j)$ denote the two-nucleon potentials between particle i and j as given by (7.21a,b). Since the total wave function is antisymmetric in i and j, the terms corresponding to i = j will automatically be eliminated when H acts on the wave function. In (7.21a,b) the tensor force potentials are known to be small, and to avoid unnecessary complications we may ignore them for the time being. Hence, we write the two-nucleon potentials as:

$$v(\vec{x}-\vec{y}) = v_1(\vec{x}-\vec{y}) + v_2(\vec{x}-\vec{y})P_\sigma + v_3(\vec{x}-\vec{y})P_\tau + v_4(\vec{x}-\vec{y})P_M \, . \tag{7.24}$$

For a uniform system it is appropriate to normalize the single-nucleon wave function inside a box of volume Ω satisfying periodic boundary conditions as it is done in Section 2. Translational invariance implies that the spatial part of the single-nucleon wave functions are given by plane waves:

$$\phi_j(\vec{x}) = \frac{1}{\sqrt{\Omega}} \exp(i\, \vec{k}\cdot\vec{x}) \, , \tag{7.25}$$

where $\hbar\vec{k} = \vec{p}$ denotes the momentum of the state. These plane wave functions are also solutions of the Hartree-Fock equations (2.25). This result accounts for the appeal and simplicity of the nuclear matter problem. The many-body Hamiltonian can be evaluated directly from these wave functions. Such is not the case with finite nuclei or atomic calculations, where one has to go through very tedious self-consistent calculations merely to generate the Hartree-Fock many-body wave function.

Together with the spin and isospin wave functions a complete description of a single-nucleon wave function is written as:

$$\psi_j(\vec{x}) = \frac{1}{\sqrt{\Omega}} \exp(i\,\vec{k}\cdot\vec{x})\chi_\sigma(j)\zeta_q(j) \;, \tag{7.26}$$

where the index j labels the jth nucleon in the system and denotes implicitly the quantum numbers of the single-nucleon state, $j=(\vec{k},\sigma,q)$. The Hartree-Fock wave function Ψ is a completely antisymmetrized function in the nucleon labels. The total energy of the system is the expectation value of the Hamiltonian evaluated from:

$$E = \int d^N v\, \Psi^* H \Psi \,. \tag{7.27}$$

Using the orthonormal properties of the single-nucleon wave functions the expectation value is given by:

$$E = \sum_{i=1}^{N} \frac{\hbar^2 k_i^2}{2m} + \frac{1}{2}\sum_{i}^{N}\sum_{j}^{N}\int d^3x d^3y\, \psi_i^+(\vec{x})\psi_j^+(\vec{y}) v(\vec{x}-\vec{y})\{\psi_i(\vec{x})\psi_j(\vec{y})-\psi_j(\vec{x})\psi_i(\vec{y})\} \;, \tag{7.28}$$

where the first summation of terms consists of the kinetic energies of the individual nucleons and the second double summations evaluate the interaction energies. The interaction energy of each pair of nucleons is usually separated into two parts:direct and exchange. The matrix elements of the direct and exchange interactions are denoted by V_D and V_E, respectively. They are given explicitly as follows:

$$V_D = \Omega^{-2}\int d^3x d^3y\, e^{-i\vec{k}\cdot\vec{x}} e^{-i\vec{k}'\cdot\vec{y}} \chi_\sigma^+(1)\chi_{\sigma'}^+(2)\zeta_q^+(1)\zeta_{q'}^+(2) \times$$
$$\times \{v_1+v_2P_\sigma+v_3P_\tau+v_4P_M\}\, e^{i\vec{k}\cdot\vec{x}} e^{i\vec{k}'\cdot\vec{y}} \chi_\sigma(1)\chi_{\sigma'}(2)\zeta_q(1)\zeta_{q'}(2) \;, \tag{7.29}$$

$$V_E = \Omega^{-2}\int d^3x d^3y\, e^{-i\vec{k}\cdot\vec{x}} e^{-i\vec{k}'\cdot\vec{y}} \chi_\sigma^+(1)\chi_{\sigma'}^+(2)\zeta_q^+(1)\zeta_{q'}^+(2) \times$$
$$\times \{v_1+v_2P_\sigma+v_3P_\tau+v_4P_M\}\, e^{i\vec{k}'\cdot\vec{x}} e^{i\vec{k}\cdot\vec{y}} \chi_{\sigma'}(1)\chi_\sigma(2)\zeta_{q'}(1)\zeta_q(2) \,. \tag{7.30}$$

The spin matrix elements are to be evaluated as follows:

$$\chi_{\sigma_1}^+(1)\chi_{\sigma_2}^+(2)\chi_{\sigma_3}(1)\chi_{\sigma_4}(2) = \left(\chi_{\sigma_1}^+(1)\chi_{\sigma_3}(1)\right)\left(\chi_{\sigma_2}^+(2)\chi_{\sigma_4}(2)\right) = \delta(\sigma_1,\sigma_3)\delta(\sigma_2,\sigma_4)$$

and

$$\chi_{\sigma_1}^+(1)\chi_{\sigma_2}^+(2)P_\sigma\chi_{\sigma_3}(1)\chi_{\sigma_4}(2) = \left(\chi_{\sigma_1}^+(1)\chi_{\sigma_4}(1)\right)\left(\chi_{\sigma_2}^+(2)\chi_{\sigma_3}(2)\right) = \delta(\sigma_1,\sigma_4)\delta(\sigma_2,\sigma_3) \tag{7.31}$$

Similarly, for the isospin matrix elements:

$$\zeta^+_{q_1}(1)\zeta^+_{q_2}(2)\zeta_{q_3}(1)\zeta_{q_4}(2) = \delta(q_1,q_3)\delta(q_2,q_4) ,$$
$$\zeta^+_{q_1}(1)\zeta^+_{q_2}(2)\; P_\tau\; \zeta_{q_3}(1)\zeta_{q_4}(2) = \delta(q_1,q_4)\delta(q_2,q_3) . \qquad (7.32)$$

Introducing also the relative coordinates $\vec{r} = \vec{x}-\vec{y}$, and the center-of-mass coordinates $\vec{R} = (\vec{x}+\vec{y})/2$, the interaction matrix elements are given by:

$$V_D = \left(\frac{1}{\Omega} \int d^3R \right) \frac{1}{\Omega} \int d^3r \{ v_1(r)+v_2(r)\delta(\sigma,\sigma')+v_3(r)\delta(q,q')-v_4(r)\delta(\sigma,\sigma')\delta(q,q')\} ,$$
$$V_E = \left(\frac{1}{\Omega} \int d^3R \right) \frac{1}{\Omega} \int d^3r\, e^{-i(\vec{k}-\vec{k}')\cdot\vec{r}} \{v_1\delta(\sigma,\sigma')\delta(q,q')+v_2\delta(q,q')+v_3\delta(\sigma,\sigma')-v_4\} .$$
$$(7.33)$$

For E to be the ground state energy of the system we need to sum up the lowest occupied momentum states. Let us denote the highest momentum reached by the system as p_F, the Fermi momentum. The summation over the states may be converted into an integral since the single-nucleon states are distributed uniformly over the momentum space:

$$\sum_i \to \frac{\Omega}{h^3} \sum_{\sigma,q}\!\!\sum \int d^3p\; \Theta(p_F-p) = \frac{\Omega}{(2\pi)^3} \sum_{\sigma,q}\!\!\sum \int d^3k\; \Theta(k_F-k) , \qquad (7.34)$$

where the Θ-function is the step-function:

$$\Theta(x) = \begin{cases} 1 & \text{for } x \geq 0 , \\ 0 & \text{for } x < 0 . \end{cases} \qquad (7.35)$$

Thus, the total kinetic energy of the system is evaluated to be:

$$\sum_i \frac{\hbar^2 k_i^2}{2m} = \frac{4\hbar^2}{2m} \frac{\Omega}{(2\pi)^3} \int_0^{k_F} 4\pi\; k^2 dk = (0.6)\; \frac{\hbar^2 k_F^2}{2m} N , \qquad (7.36)$$

where

$$N = 4 \frac{\Omega}{(2\pi)^3} \frac{4\pi\; k_F^3}{3}$$

is the total number of nucleons in the system. The interaction energies are given by:

$$\frac{1}{2} \sum_{i,j} (V_D-V_E) = \frac{1}{2} \sum_{\sigma,q} \sum_{\sigma',q'} \frac{\Omega^2}{(2\pi)^6} \int d^3k d^3k'\; \Theta(k_F-k)\Theta(k_F-k')\{V_D-V_E\}$$
$$= \frac{1}{2} \frac{\Omega}{(2\pi)^6} \int d^3k d^3k' \Theta(k_F-k)\Theta(k_F-k') d^3r \{ (16v_1+8v_2+8v_3-4v_4) + $$
$$- e^{-i(\vec{k}-\vec{k}')\cdot\vec{r}} (4v_1+8v_2+8v_3-16v_4)\} .$$

This expression may further be reduced by making use of the expansion:

$$e^{i\vec{k}\cdot\vec{r}} = \sum_{\ell=0}^{\infty} (2\ell+1)\, i^{\ell}\, j_{\ell}(kr)\, P_{\ell}(\hat{k}\cdot\hat{r}) , \tag{7.37}$$

where $j_{\ell}(x)$ are the spherical Bessel functions, $P_{\ell}(x)$ are the Legendre polynomials (see Appdendix) and $\hat{k}$ and $\hat{r}$ are unit vectors. The momentum space integrals can be evaluated as follows:

$$\int d^3k\, \Theta(k_F-k)\, e^{i\vec{k}\cdot\vec{r}} = 4\pi \int_0^{k_F} k^2 dk\, j_0(kr) = \frac{4\pi\, k_F^3}{3}\, \frac{3\, j_1(k_F r)}{k_F r} . \tag{7.38}$$

Finally, the energy per particle is given by:

$$\frac{E}{N} = (0.6)\, \frac{\hbar^2 k_F^2}{2m} + \frac{k_F^3}{12\pi^2} \Big\{ \int d^3r (4v_1+2v_2+2v_3-v_4) + \int d^3r (4v_4-2v_3-2v_2-v_1) \Big(\frac{3j_1(k_F r)}{k_F r}\Big)^2 \Big\} \tag{7.39}$$

In the Hartree-Fock approximation each nucleon in nuclear matter experiences an overall potential, which we shall refer to as a <u>single-particle potential</u>. For a nucleon in the state of spin σ, isospin q and momentum $\vec{k}$, the single-particle potential is given by:

$$U_{\sigma,q}(\vec{k}) = \sum_{\sigma',q'} \frac{\Omega}{(2\pi)^3} \int d^3k' \Theta(k_F-k') (\, V_D - V_E \,) =$$

$$= \frac{1}{(2\pi)^3} \int d^3k' \Theta(k_F-k') d^3r \,\{ (16v_1+8v_2+8v_3-4v_4) + - e^{-i(\vec{k}-\vec{k}')\cdot\vec{r}} (4v_1+8v_2+8v_3-16v_4) \} =$$

$$= \frac{k_F^3}{6\pi^2} \Big\{ \int d^3r\, (4v_1+2v_2+2v_3-v_4) + \int d^3r\, (4v_4-2v_3-2v_2-v_1)\, j_0(kr) \Big(\frac{3\, j_1(k_F r)}{k_F r}\Big) \Big\} . \tag{7.40}$$

It turns out that $U_{\sigma,q} \to U$ is independent of spin σ and isospin q. For small momenta, $j_0(kr)$ may be expanded into,

$$j_0(kr) = 1 + k^2 r^2/6 + \ldots . \tag{7.41}$$

and (7.40) is given approximately by:

$$U_{\sigma,q} \to U(k) = \frac{k_F^3}{6\pi^2} \Big\{ \int d^3r (4v_1+2v_2+2v_3-v_4) + \int d^3r (4v_4-2v_3-2v_2-v_1) \frac{3j_1(k_F r)}{k_F r} - \Big(\frac{k^2}{6}\Big) \int d^3r\, r^2 (4v_4-2v_3-2v_2-v_1) \frac{3j_1(k_F r)}{k_F r} + \ldots \Big\}$$

$$\equiv U_0 + \frac{\hbar^2 k^2}{2m} U_1 + \ldots . \tag{7.42}$$

leading to a parabolic single-particle potential at small k. In the opposite limit of large k, $j_0(kr)$ oscillates rapidly and the exchange integral vanishes, thus,

$$U(k) \underset{k\to\infty}{\rightarrow} \frac{k_F^3}{6\pi^2} \int d^3r(4v_1+2v_2+2v_3-v_4) \ . \tag{7.43}$$

A sketch of U(k) is shown schematically in Figure 7-1.

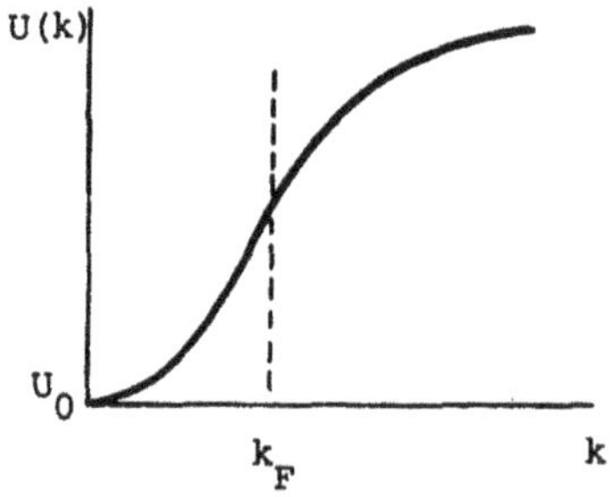

Figure 7-1. Sketch of the single-particle potential U(k) in (7.40).

The Hartree-Fock method applied here is also referred to as the nuclear independent-particle model. It does not predict the saturation properties of nuclear matter if realistic nuclear potentials obtainable from nucleon-nucleon scattering experiments are employed. To overcome this defect, two different approaches may be taken. One is to introduce effective nuclear potentials which are designed for the independent-particle model and are qualitatively correct but with adjustable parameters so as to reproduce nuclear matter properties. The other is to improve on the independent-particle model to include particle correlation effects into the model. These approaches will be described separately in subsequent sections.

We have not exhausted the possible forms of the nuclear potential function. Some may depend on the momenta of the particles, which are denoted $\vec{p}_1$ and $\vec{p}_2$, if velocity-dependent potentials are considered. Just as translational invariance of the potential function restricts the functional dependence to the relative coordinates $\vec{r} = (\vec{x}_1-\vec{x}_2)$, the Galilean invariance of the potential function will also restrict it to depend on the relative momentum $\vec{p} = (\vec{p}_1-\vec{p}_2)$. In general, the potential may be any arbitrary functions of the following scalars: $\vec{r}^2$, $\vec{p}^2$, and $\vec{L}^2$, where $\vec{L} = \vec{r}\times\vec{p}$ with $\vec{p} = -i\hbar\vec{\nabla}$. In addition there are spin dependent terms. A term linear in spin is the spin-orbit term $(\vec{S}\cdot\vec{L})$ which plays an important role in the nuclear shell model. On the other hand, terms like $(\vec{S}\cdot\vec{r})$ and $(\vec{S}\cdot\vec{p})$ are forbidden by space-reflection invariance, which requires that $v(\vec{r},\vec{p}) = v(-\vec{r},-\vec{p})$. In addition to the tensor operator S_{12} a second operator bilinear in spin is:

$$Q_{12} = \frac{1}{2}\{(\vec{s}_1\cdot\vec{L})(\vec{s}_2\cdot\vec{L}) + (\vec{s}_2\cdot\vec{L})(\vec{s}_1\cdot\vec{L})\} \qquad (7.44)$$

which is symmetrized with respect to nucleons 1 and 2. The problem of constructing the most general potential function consistent with accepted invariance principles has been discussed by Eisenbud and Wigner (1941) and by Okubo and Marshak (1958). The allowed expression for a potential function may be too general for practical purposes. In subsequent studies we shall investigate specific potentials and carry out all evaluations in some detail.

References

Bethe, H.A. (1971). Ann. Rev. Nucl. Sci. 21, 93.

Brueckner,K.A., Gammel,J.A. and Thaler, R.M. (1958). Phys. Rev. 109, 1023.

Eisenbud, L. and Wigner, E. (1941). Proc. Nat. Acad. Sci. 27, 281.

Gammel, J.A. and Thaler, R. (1957). Phys. Rev. 107, 1337.

Okubo, S. and Marshak, R.E. (1958). Ann. Phys. 4, 166.

Bibliography

Barrett,R.C. and Jackson, D.F. (1979). Nuclear Sizes and Structures, Clarendon Press, Oxford.

Brink, D.M. (1965). Nuclear Forces, Pergamon Press, New York.

Lomon, E.L. and Feshbach, H. (1967). Rev. Mod. Phys. 39, 611.

Preston, M.A. and Bhaduri, R.K. (1975). Structure of the Nucleus, Addison-Wesley,Reading, Mass.

Signell, P. (1972). In The Two-Body Force in Nuclei, (Ed. Austin, S.M. and Crawley, G.M.), p. 9. Plenum Press, New York.

Wilson, R. (1963). The Nucleon-Nucleon Interaction, Experimental and Phenomenological Aspects, Wiley Interscience, New York.

8. Nuclear Effective Interactions

Nuclear physics calculations in the Hartree-Fock approximation can be made much more successful if instead of the true two-nucleon interactions, one uses phenomenological effective interactions in the calculations. However, the fact that such effective interactions are tied in with a specific method of computation underscores its limitations. They are not to be considered as fundamental entities, but they nevertheless provide a useful vehicle for interpolating known facts in much the same way as the semi-empirical mass formula or specific nuclear models. Many nuclear physics calculations will have to rely on such semi-empirical schemes for now.

The effective interactions are to be used in the framework of the Hartree-Fock method. They are required to fit the saturation properties of nuclear matter as well as the static properties of finite nuclei. We have mentioned before that nuclear matter shows a saturation density. This can be accomplished by effective interactions which are density dependent so that interactions become less and less attractive (and eventually repulsive) as the density increases, or which are velocity dependent so that as the Fermi energies of the degenerate nucleons become large as density increases, there will be a reduction in attraction. Of course, effective interactions may contain both mechanisms. We shall describe here effective interactions given by the Skyrme potentials.

The Skyrme potentials (Skyrme 1956, 1959) have gained a great deal of popularity in the recent years due to much seeming success in employing them for the study of finite nuclei, and to the simplicity they offer in an actual calculation. They consist of both velocity dependent as well as density dependent parts. The Skyrme potentials consist of a two-body part which is given as follows:

$$v(\vec{x}_1-\vec{x}_2) = t_0(1+x_0P_\sigma)\delta(\vec{x}_1-\vec{x}_2) + \frac{1}{2}t_1\{\delta(\vec{x}_1-\vec{x}_2)\vec{K}^2 + \overleftarrow{K}'^2\delta(\vec{x}_1-\vec{x}_2)\} + $$
$$+ t_2\overleftarrow{K}'\cdot\delta(\vec{x}_1-\vec{x}_2)\vec{K} + iW_0(\vec{\sigma}_1+\vec{\sigma}_2)\cdot\overleftarrow{K}'\times\delta(\vec{x}_1-\vec{x}_2)\vec{K}\ , \qquad (8.1)$$

where t_0, t_1, t_2, x_0, and W_0 are parameters, $\vec{K}$ denotes the operator $(\vec{\nabla}_1-\vec{\nabla}_2)/2i$ acting onto the right and $\overleftarrow{K}'$ denotes the operator $-(\overleftarrow{\nabla}_1-\overleftarrow{\nabla}_2)/2i$

acting onto the left. The Skyrme potentials contains also a three-body term given by:

$$v_3(\vec{x}_1,\vec{x}_2,\vec{x}_3) = t_3\,\delta(\vec{x}_1-\vec{x}_2)\delta(\vec{x}_2-\vec{x}_3)\ . \tag{8.2}$$

The gradient operators acting on the particle wave functions provide the velocity dependent forces. The simplicity of these potentials lies in the fact that their spatial dependences are given by δ-functions. They provide zero-range forces. The three-body terms as we shall see is equivalent to a two-body density dependent interaction for spherically symmetric situations.

Let us illustrate the meaning of the Skyrme potentials by repeating the nuclear matter problem worked out in the previous section for static potentials. The single-nucleon wave functions will again be given by (7.26) and the total energy of the system by (7.28). We shall calculate each of the interaction terms separately.

(1) terms proportional to t_0:

$$V_D = \frac{t_0}{\Omega^2}\int d^3x_1 d^3x_2\ e^{-i\vec{k}_1\cdot\vec{x}_1}\,e^{-i\vec{k}_2\cdot\vec{x}_2}\,\delta(\vec{x}_1-\vec{x}_2)\,e^{i\vec{k}_1\cdot\vec{x}_1}\,e^{i\vec{k}_2\cdot\vec{x}_2}\ \times$$

$$\times\ \{(\chi^+_{\sigma_1}\chi_{\sigma_1})(\chi^+_{\sigma_2}\chi_{\sigma_2}) + x_0(\chi^+_{\sigma_1}\chi_{\sigma_2})(\chi^+_{\sigma_2}\chi_{\sigma_1})\}\ (\zeta^+_{q_1}\zeta_{q_1})(\zeta^+_{q_2}\zeta_{q_2}) =$$

$$= \frac{1}{\Omega}\int d^3r\,\frac{1}{\Omega}t_0\ \{1 + x_0\ \delta(\sigma_1,\sigma_2)\}\ ,$$

$$V_E = \frac{t_0}{\Omega^2}\int d^3x_1 d^3x_2\ e^{-i\vec{k}_1\cdot\vec{x}_1}\,e^{-i\vec{k}_2\cdot\vec{x}_2}\,\delta(\vec{x}_1-\vec{x}_2)\,e^{i\vec{k}_2\cdot\vec{x}_1}\,e^{i\vec{k}_1\cdot\vec{x}_2}\ \times$$

$$\times\ \{\ (\chi^+_{\sigma_1}\chi_{\sigma_2})(\chi^+_{\sigma_2}\chi_{\sigma_1}) + x_0(\chi^+_{\sigma_1}\chi_{\sigma_1})(\chi^+_{\sigma_2}\chi_{\sigma_2})\}(\zeta^+_{q_1}\zeta_{q_2})(\zeta^+_{q_2}\zeta_{q_1})$$

$$= \frac{1}{\Omega}\int d^3r\,\frac{1}{\Omega}t_0\ \{\ \delta(\sigma_1,\sigma_2) + x_0\}\ \delta(q_1,q_2)\ ,$$

and,

$$V_D - V_E = \frac{1}{\Omega}\Big(\ t_0\{1 - \delta(\sigma_1,\sigma_2)\delta(q_1,q_2)\} + t_0x_0\{\delta(\sigma_1,\sigma_2)-\delta(q_1,q_2)\}\Big)\ . \tag{8.3}$$

Its contribution to the total interaction energy is given by:

$$U_0 = \frac{1}{2}\sum_{\vec{k}_1,\sigma_1,q_1}\ \sum_{\vec{k}_2,\sigma_2,q_2}\{\ V_D - V_E\ \} = \frac{1}{2}\sum_{\sigma_1,q_1}\ \sum_{\sigma_2 q_2}\frac{\Omega^2}{(2\pi)^6}\int d^3k_1 d^3k_2\ \{V_D-V_E\}$$

$$= t_0\ \Omega\,\frac{k_F^6}{6\pi^4} = \frac{3}{8}\,t_0\ n_B\ A\ , \tag{8.4}$$

where $A = 2\Omega\, k_F^3/(3\pi^2)$ is the total number of nucleons in the system as given by (7.36), and $n_B = (A/\Omega)$ is the nucleon number density. We use here the symbol A instead of N for the total number of nucleons as it is customary in nuclear physics.

(2) terms proportional to t_1: (spin and isospin factors may be worked out as above and will not be repeated here)

$$V_D = \frac{t_1}{2\Omega^2}\int d^3x_1 d^3x_2\,\{e^{-i\vec{k}_1\cdot\vec{x}_1} e^{-i\vec{k}_2\cdot\vec{x}_2}\delta(\vec{x}_1-\vec{x}_2)\left(-\frac{1}{4}(\vec{\nabla}_1-\vec{\nabla}_2)\cdot(\vec{\nabla}_1-\vec{\nabla}_2)e^{i\vec{k}_1\cdot\vec{x}_1} e^{i\vec{k}_2\cdot\vec{x}_2}\right\} +$$

$$- \left(-\frac{1}{4}(\vec{\nabla}_1-\vec{\nabla}_2)\cdot(\vec{\nabla}_1-\vec{\nabla}_2)e^{-i\vec{k}_1\cdot\vec{x}_1} e^{-i\vec{k}_2\cdot\vec{x}_2}\right\}\delta(\vec{x}_1-\vec{x}_2)e^{i\vec{k}_1\cdot\vec{x}_1} e^{i\vec{k}_2\cdot\vec{x}_2}\}$$

$$= \frac{t_1}{2\Omega} \times \frac{1}{2}(\vec{k}_1^2 + \vec{k}_2^2 - 2\vec{k}_1\cdot\vec{k}_2)\ .$$

Similarly for V_E, so that,

$$V_D - V_E = \frac{t_1}{4\Omega} \times (\vec{k}_1^2 + \vec{k}_2^2 - 2\vec{k}_1\cdot\vec{k}_2)\{\,1 - \delta(\sigma_1,\sigma_2)\delta(q_1,q_2)\}. \qquad (8.5)$$

Its contribution to the interaction energy is:

$$U_1 = \left(\frac{9}{80}\right) t_1\, k_F^2\, n_B\, A\ . \qquad (8.6)$$

(3) terms proportional to t_2:

$$V_D = \frac{t_2}{\Omega^2}\int d^3x_1 d^3x_2\{-\frac{1}{2i}(\vec{\nabla}_1-\vec{\nabla}_2)e^{-i\vec{k}_1\cdot\vec{x}_1} e^{-i\vec{k}_2\cdot\vec{x}_2}\}\delta(\vec{x}_1-\vec{x}_2) \times$$

$$\times\{\frac{1}{2i}(\vec{\nabla}_1-\vec{\nabla}_2)e^{i\vec{k}_1\cdot\vec{x}_1} e^{i\vec{k}_2\cdot\vec{x}_2}\}$$

$$= \frac{t_2}{\Omega} \times \frac{1}{4}(\vec{k}_1^2 + \vec{k}_2^2 - 2\vec{k}_1\cdot\vec{k}_2)\ ,$$

and,

$$V_D - V_E = \frac{t_2}{\Omega} \times \frac{1}{4}(\vec{k}_1^2 + \vec{k}_2^2 - 2\vec{k}_1\cdot\vec{k}_2)\{1 + \delta(\sigma_1,\sigma_2)\delta(q_1,q_2)\}\ . \qquad (8.7)$$

Its contribution to the interaction energy is:

$$U_2 = \left(\frac{3}{16}\right) t_2\, k_F^2\, n_B\, A\ . \qquad (8.8)$$

(4) terms proportional to W_0:

$$V_D = \frac{W_0}{\Omega^2}\int d^3r\,\{(\chi^+_{\sigma_1}\vec{\sigma}\,\chi_{\sigma_1})(\chi^+_{\sigma_2}\chi_{\sigma_2}) + (\chi^+_{\sigma_1}\chi_{\sigma_1})(\chi^+_{\sigma_2}\vec{\sigma}\chi_{\sigma_2})\}\cdot(\vec{k}_1-\vec{k}_2)\times(\vec{k}_1-\vec{k}_2) = 0\ ,$$

$$V_D - V_E = 0 \qquad (8.9)$$

The interaction energy involving the three-body potential can be written as:

$$U_3 = \frac{1}{3!}\sum_{i,j,k} \int d^3x_1 d^3x_2 d^3x_3 \ \psi_i^+(\vec{x}_1)\psi_j^+(\vec{x}_2)\psi_k^+(\vec{x}_3) \ v_3 \ \Psi_{ijk} \ \Omega^{-3} \tag{8.10}$$

where Ψ_{ijk} is the three-particle Slater determinant:

$$\Psi_{ijk} = \begin{vmatrix} \psi_i(\vec{x}_1) & \psi_i(\vec{x}_2) & \psi_i(\vec{x}_3) \\ \psi_j(\vec{x}_1) & \psi_j(\vec{x}_2) & \psi_j(\vec{x}_3) \\ \psi_k(\vec{x}_1) & \psi_k(\vec{x}_2) & \psi_k(\vec{x}_3) \end{vmatrix} . \tag{8.11}$$

Due to the simplicity of v_3, U_3 can readily be evaluated:

$$\begin{aligned} U_3 &= \frac{t_3}{6}\sum_{\vec{k}_1\sigma_1q_1}\sum_{\vec{k}_2\sigma_2q_2}\sum_{\vec{k}_3\sigma_3q_3}\Omega^{-2}\int d^3r\{1 - \delta(\sigma_1,\sigma_2)\delta(q_1,q_2) - \delta(\sigma_2,\sigma_3)\delta(q_2,q_3) + \\ &\qquad - \delta(\sigma_1,\sigma_3)\delta(q_1,q_3) + 2\delta(\sigma_1,\sigma_2)\delta(\sigma_2,\sigma_3)\delta(q_1,q_2)\delta(q_2,q_3)\} \\ &= \frac{t_3}{6}\left(\frac{k_F^3}{6\pi^2}\right)^3 \Omega\{4\times4\times4 - 4\times4 - 4\times4 - 4\times4 + 2\times4\} \\ &= \frac{t_3}{16} n_B^2 A . \end{aligned} \tag{8.12}$$

Finally, together with the kinetic energy term, the average energy per nucleon is given by:

$$\frac{E}{A} = \frac{3}{5}\left(\frac{\hbar^2 k_F^2}{2m}\right) + n_B\left(\frac{3t_0}{8} + \frac{9t_1}{80}k_F^2 + \frac{3t_2}{16}k_F^2\right) + \frac{t_3}{16}n_B^2 . \tag{8.13}$$

The parameters of the Skyrme potentials are designed to fit first of all the average energy per nucleon in nuclear matter. However, different investigations may use different saturation densities for nuclear matter, and the Skyrme potential parameters being phenomenological should only be used for that particular saturation Fermi momentum k_{FO} intended. For example, the original Skyrme parameters (Skyrme 1959) are designed for k_{FO}= 1.37 fm^{-1}, and E/A = -17.04 MeV. Subsequently, Vautherin and Brink (1970) made extensive computations with the Skyrme potentials for finite nuclei. They found that Skyrme's parameters gave too small radii for heavy nuclei. Consequently, they generated two new sets of parameters: one for k_{FO}= 1.32 fm^{-1} and the other for k_{FO}= 1.30 fm^{-1}, with both of these producing an average energy per particle of E/A= -16 MeV. In Table 8-1, we summarize some of the proposed parameters

Table 8-1. List of parameters proposed for the Skyrme potentials.

	t_0 (MeV-fm^3)	t_1 (MeV-fm^5)	t_2 (MeV-fm^5)	t_3 (MeV-fm^6)	x_0	W_0 (MeV-fm^5)
Skyrme	-1072	461	-40	8027		
SI	-1057.3	235.9	-100.0	14463.5	0.56	120.0
SII	-1169.9	586.6	-27.1	9331.1	0.34	105.0
SIII	-1128.75	395.0	-95.0	14000.0	0.45	120.0
SIV	-1205.6	765.0	35.0	5000.0	0.05	150.0
SV	-1248.29	970.56	107.22	0	-0.17	150.0
SVI	-1101.81	271.67	-138.33	17000.0	0.583	115.0
SKM	-2645.0	385.0	-120.0	15595.0	0.09	
RBP	-1057.3	235.9	-100.0	14463.5	0.2885	120.0

for the Skyrme potentials. We designate the original Skyrme's parameters by "Skyrme" and those proposed by Vautherin and Brink by SI and SII. More recently, several other new sets of the Skyrme potential parameters also appeared. They are listed as SIII, SIV, SV and SVI in Table 8-1 (Beiner et al. 1975). They exhibit the large degrees of freedom which remain after imposing the saturation properties of nuclear matter. When applied to even-even nuclei, different sets of parameters predict somewhat different total binding energies and charge radii for these nuclei, but they show good overall agreements with experimental data. Thus, Hartree-Fock calculations with the Skyrme potentials should be a useful tool. In Figure 8-1, the average per particle energies near the

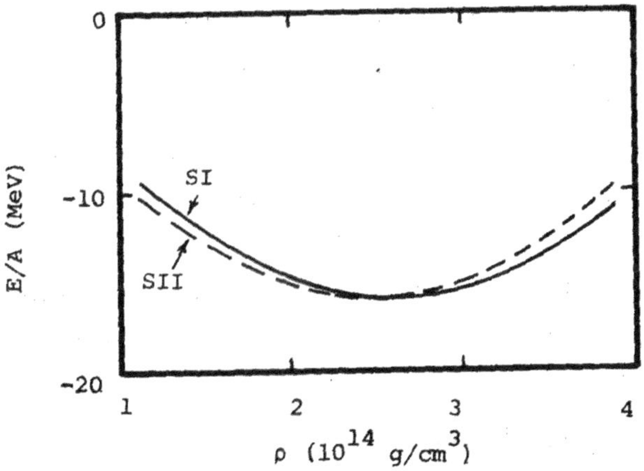

Figure 8-1. Average energy per nucleon in symmetric nuclear matter determined by Skyrme potentials with SI and SII parameters.

saturation density as predicted by the SI and SII parameters are illustrated. Saturation density occurs of course at the bottom of the bell-shaped curve.

All sets of the Skyrme parameters mentioned above, however, give too high a compression modulus for nuclear matter. The compression modulus K for nuclear matter is defined to be:

$$K = k_{F0}^2 \left.\frac{d^2(E/A)}{dk_F^2}\right|_{k_{F0}} = 9\, n_0^2 \left.\frac{d^2(E/A)}{dn_B^2}\right|_{n_0} . \tag{8.14}$$

Experimentally, it has been estimated recently to be K = 210±30 MeV (Blaizot 1980). It is calculated from the Skyrme potentials to be:

$$K = \frac{6}{5}\frac{\hbar^2}{2m} k_{F0}^2 + \left(\frac{9}{4} t_0 + \frac{9}{4} t_1 k_{F0}^2 + \frac{15}{4} t_2 k_{F0}^2 \right) n_B + \frac{15}{8} t_3 n_B^2 . \tag{8.15}$$

Making use of (8.13) and the saturation condition:

$$\left.\frac{d(E/A)}{dk_F}\right|_{k_{F0}} = 0 , \tag{8.16}$$

the compression modulus K is found to depend on just one of the Skyrme parameters:

$$K = -15(E/A) + \frac{9}{5}\frac{\hbar^2}{2m} k_{F0}^2 + \frac{1}{12\pi^4} t_3 k_{F0}^6 . \tag{8.17}$$

This means that for any set of Skyrme parameters possessing a positive value of t_3 and predicting reasonable values of E/A and k_{F0}, it must have a compression modulus larger than 300 MeV as given by (8.17). Such a value for K is high and there seems to be a dilemma here. One way out of this dilemma is to reinterpret the t_3 term. It is originally constructed as a three-body interaction term, but in a Hartree-Fock calculation, it is no more than a two-body density-dependent interaction term, since (8.12) shows that U_3 contains an extra power of n_B as compared to the other interaction terms. This shows that the t_3 term may be reinterpreted as a two-body interaction term with a linear density dependence. Once such a reinterpretation is accepted, one may allow the power of the density dependence to be phenomenological by writing v_3 of (8.2) as a two-body potential in the form:

$$v_3(\vec{x}_1,\vec{x}_2) = \frac{1}{6} t_3 \left\{ n_B\!\left(\frac{\vec{x}_1+\vec{x}_2}{2}\right) \right\}^{\alpha} \delta(\vec{x}_1-\vec{x}_2) . \tag{8.18}$$

One such a modified set of Skyrme parameters, designated SKM (Krivine, Treiner, Bohigas 1980) with $\alpha = (1/6)$ is listed in Table 8-1. The compression modulus K for nuclear matter predicted by SKM potentials is reduced to an acceptable value of K = 213 MeV. There are also other similarly modified sets. Some include an additional parameter by multiplying v_3 by $(1 + x_3 P_\sigma)$ where x_3 is an adjustable parameter and P_σ is the spin exchange operator. A modified SI interaction denoted in Table 8-1 as RBP (Ravenhall, Benett, Pethick 1971) has the same SI parameters and it is different from the SI interaction by having a v_3 term given by (8.18) with $\alpha = 1$, which is further multiplied by $(1 + 0.2257P_\sigma)$.

The average per particle energy computed from (8.13) is for nuclear matter which consists of exactly half protons and half neutrons (symmetric nuclear matter). Since it is a bound system, the pressure is negative when the density of the system is below the saturation density. Under special conditions, as in the case of a supernova process, nature favors matter which is not symmetric in neutron-proton distribution. In this case the Skyrme potentials may still be used to study the equation of state of such matter. As in Section 2, let us introduce the electron fraction Y_e, which is the average number of electrons per nucleon in the system. Due to charge neutrality of the system, it is also the proton fraction of the system. In terms of Y_e the proton number density is given by $n_p = Y_e n_B$ and the neutron number density by $n_n = (1-Y_e)n_B$. Let us now consider situations where $Y_e \neq 0.5$ and reevalue U_0, U_1, U_2, and U_3 of (8.4) to (8.12) for $n_p \neq n_n$. They are:

$$U_0 = \Omega^{-1}(t_0/2) \sum_{\vec{k}_1\sigma_1 q_1} \sum_{k_2\sigma_2 q_2} \{ \left(1 - \delta(\sigma_1,\sigma_2)\delta(q_1,q_2)\right) + x_0\left(\delta(\sigma_1,\sigma_2)\ \delta(q_1,q_2)\right) \}$$

$$= \Omega^{-1}\ t_0 \sum_{\vec{k}_1 q_1} \sum_{\vec{k}_2 q_2} \{ \left(2 - \delta(q_1,q_2)\right) + x_0 \left(1 - 2\delta(q_1,q_2)\right) \}$$

$$= \Omega\ t_0 \frac{(4\pi)^2}{(2\pi)^6} \{ (1-x_0)\left(\int_0^{k_{Fp}} k_1^2 dk_1 \int_0^{k_{Fp}} k_2^2 dk_2 + \int_0^{k_{Fn}} k_1^2 dk_1 \int_0^{k_{Fn}} k_2^2 dk_2 \right) +$$

$$+ (2+x_0)\left(\int_0^{k_{Fp}} k_1^2 dk_1 \int_0^{k_{Fn}} k_2^2 dk_2 + \int_0^{k_{Fn}} k_1^2 dk_1 \int_0^{k_{Fp}} k_2^2 dk_2 \right) \}$$

$$= \Omega\ (t_0/2)\quad (\ 1 + \frac{1}{2}x_0)(n_p+n_n)^2 - (\ \frac{1}{2} + x_0)(n_p^2+n_n^2)$$

$$= (t_0/2)\{\ (1 + \frac{1}{2}x_0) - (\frac{1}{2} + x_0)\left(Y_e^2 + (1-Y_e)^2\right)\}\ n_B A\ . \tag{8.19}$$

$$U_1+U_2 = \Omega^{-1} \sum_{\vec{k}_1 q_1} \sum_{\vec{k}_2 q_2} (k_1^2+k_2^2+2\vec{k}_1\cdot\vec{k}_2)\{(t_1+t_2) + \frac{1}{2}(t_2-t_1)\delta(q_1,q_2)\}$$

$$= \frac{1}{4}(t_1+t_2)(\tau_p+\tau_n)A + \frac{1}{8}(t_2-t_1)\{\tau_p Y_e+\tau_n(1-Y_e)\}A\ . \tag{8.20}$$

where

$$\tau_a = \frac{3}{5}\, k_{Fa}^2 n_a \qquad (\text{for } a{=}n,p)\,. \tag{8.21}$$

For v_3 given by (8.2), U_3 is evaluated to be:

$$U_3 = \frac{1}{4}\, t_3\, n_p n_n A\ , \tag{8.22}$$

and for v_3 given by (8.18), U_3' (a prime is added to distinguish the two) is evaluated to be:

$$U_3' = \frac{1}{12}\, n_B^{\alpha-1}\ \{(1+\frac{1}{2}x_3)n_B^2 - (\frac{1}{2} + x_3)(n_p^2+n_n^2)\}\ A\ . \tag{8.23}$$

which is evaluated for the parametrization with $x_3 \neq 0$. Even though v_3 given by (8.2) and (8.18) predict the same nuclear matter properties they have very different isospin contents and this fact is made apparent in U_3 and U_3'. If we look at the case of a pure neutron matter, where $n_p = 0$, then $U_3 = 0$ while U_3' is finite. The fact that U_3 vanishes in the limit $n_p = 0$ does not seem reasonable,and as a consequence when the Skyrme potentials are used to evaluated the equation of state for pure neutron matter, the SI parameters give negative pressure in the density region near the nuclear matter density,which is again quite unreasonable. Hence,for these reasons,it is advisable to employ the SKM and RBP parametrizations for dense matter studies.

The average energy per nucleon with the U_3' term is given by:

$$\begin{aligned} n_B(E/A) = {} & \frac{\hbar^2}{2m}(\tau_n+\tau_p) + \frac{1}{2}\, t_0\ \{(1+\frac{1}{2}x_0)(n_n+n_p)^2 - (\frac{1}{2} + x_0)(n_n^2+n_p^2)\} + \\ & + \frac{1}{4}\ (t_2+t_1)(\tau_n+\tau_p)(n_n+n_p) + \frac{1}{8}\ (t_2-t_1)(\tau_n n_n+\tau_p n_p) + \\ & + \frac{1}{12}\, t_3(n_n+n_p)^\alpha\ \{\frac{1}{2}(1-x_3)(n_n+n_p)^2 + (1+2x_3)n_n n_p\} \end{aligned} \tag{8.24}$$

The pressure is evaluated to be:

$$P = -\left.\frac{\partial E}{\partial \Omega}\right|_A = \Omega^{-1}\frac{1}{3}\left(k_{Fp}\frac{\partial E}{\partial k_{Fp}} + k_{Fn}\frac{\partial E}{\partial k_{Fn}} \right)$$

$$= \frac{2}{3}\frac{\hbar^2}{2m}(\tau_n+\tau_p) + \frac{1}{2}t_0\{(1+\frac{1}{2}x_0)(n_n+n_p)^2 - (\frac{1}{2}+x_0)(n_n^2+n_p^2)\} +$$

$$+ \frac{5}{12}(t_1+t_2)(\tau_n+\tau_p)(n_n+n_p) + \frac{5}{24}(t_2-t_1)(\tau_n n_n+\tau_p n_p) +$$

$$+ \frac{(\alpha+1)}{12}t_3(n_n+n_p)^\alpha\{\frac{1}{2}(1-x_3)(n_n+n_p)^2 + (1+2x_3)n_n n_p\}, \quad (8.25)$$

and its mass density, $\rho = m_p n_B$.

The equations of state for pure neutron matter at subnuclear densities are computed for the Skyrme potentials with the SKM and RBP parameters. They are plotted in Figure 8-2. For comparison we have added the free neutron equation of state, which is computed from (8.25) by setting $t_0=t_1=t_2=t_3=0$.

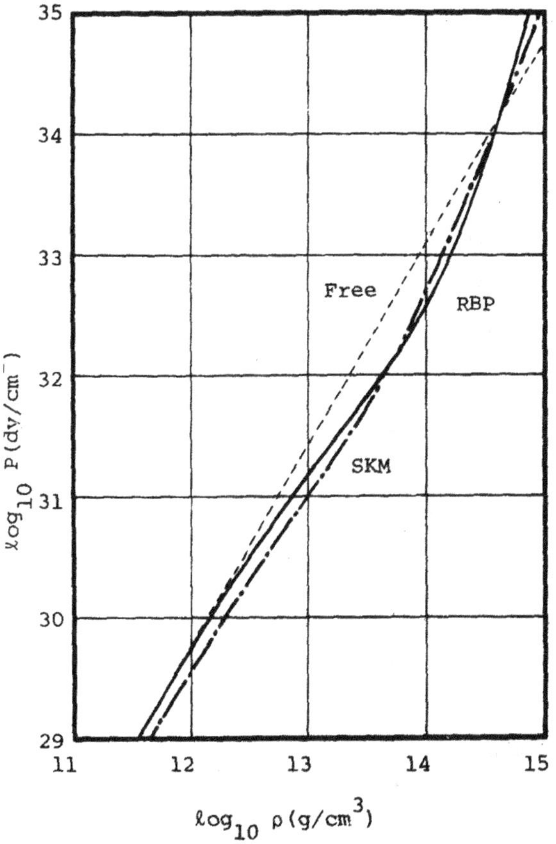

Figure 8-2. Equations of state for pure neutron matter computed from Skyrme potentials with SKM (dot-dashed lines) and RBP (solid line) parameters. Free neutron curve is in dotted lines.

For $Y_e \neq 0$, the pressure generated by the nucleons may be negative, but the presence of electrons would add to the system a positive pressure. It is instructive to look at the equation of state whose total pressure is due to the nucleons and degenerate electrons with electrostatic interaction neglected. We may take the extreme relativistic approximation for the electron pressure, so that:

$$P = P_s + P_e \quad , \tag{8.26}$$

where P_s is given by (8.25), and:

$$P_e = (3/\pi)^{1/3}(hc/8)(n_p)^{4/3} \quad . \tag{8.27}$$

Equations of state computed from (8.26) with various electron fractions Y_e are shown in Figure 8-3. They are computed for the SKM parameters. In general, the nucleon pressure is becoming more and more negative at subnuclear densities as Y_e increases from zero to 0.5. On the other hand, the increase in electron density for high Y_e substance is suffi-

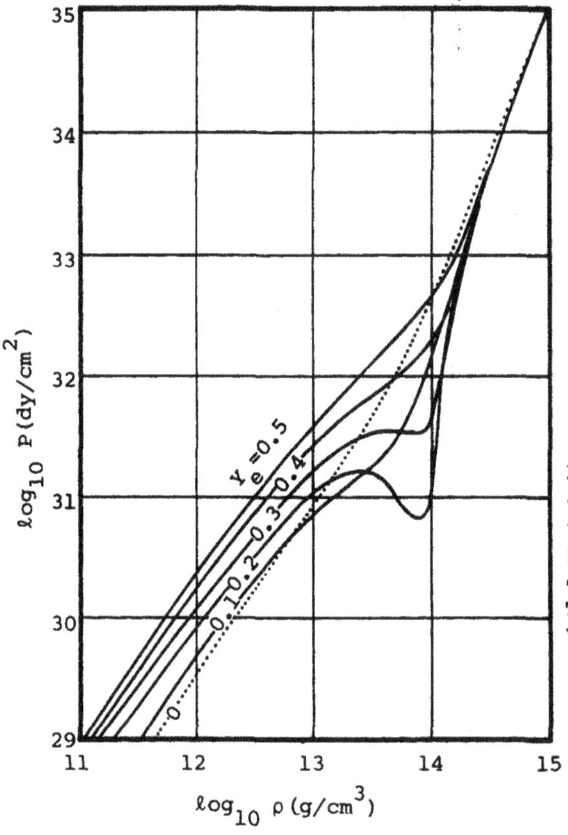

Figure 8-3. Equations of computed from Skyrme potentials with SKM parameters for systems with different electron fractions, Y_e. The pure neutron curve, Y_e=0 is in dotted lines.

cient to compensate for the loss of pressure. The more critical cases are therefore those for Y_e=0.2 or 0.3, which exhibit dips in the equations of state as shown.

An equation of state which shows a dip structure implies the possible existence of a phase transition in the system. In other words, matter at two different densities gives rise to the same pressure, and these two phases of different densities may therefore coexist. The situation is usually such that the high density phase condenses into nuclei while the low density phase occupies the region outside of the nuclei. The surface of the nuclei is subjected to a surface tension and there will be a surface energy to be associated with the nuclei. As matter density increases these nuclei also grow in size and becoming more and more neutron rich. The nucleons outside the nuclei are sometimes called the "dripped neutrons", since they only appear after the nuclei have grown into certain size and it seems that they are dripping out of the nuclei. They are mostly neutrons with only a small admixture of protons, and thus the name "dripped neutrons". An alternative terminology in describing these two phases is to refer to the dense portion as the liquid phase and the less dense portion as the vapor phase, since these two phases exhibit very different compressibilities as in the case of liquid and vapor. At nuclear density, the entire system will be in the liquid phase. At lower densities, the liquid phase condenses into droplets with the size of the droplets determined by the surface energy, and then at approximately half of the nuclear density, the liquid phase will fill the space leaving the vapor phase to exist as bubbles inside the liquid. The change over takes place in order to economize on the surface energy. A more detailed description of these situations will be given in the next section.

References

Beiner, M., Flocard, H., Nguyen Van Giai, and Quentin, P. (1975). Nucl. Phys. A238, 29.

Blaizot, J.P. (1980). Phys. Reports 64, 171.

Krivine, H., Treiner, J, and Bohigas, O. (1980). Nucl. Phys. A336, 155.

Ravenhall, D.G., Bennett, C.D., Pethick C.J. (1971). Phys. Rev. Lett. 28, 978.

Skyrme,T.H.R. (1956). Phil. Mag. 1, 1043.

Skyrme, T.H.R. (1959). Nucl. Phys. 9, 615.

Vautherin, D. and Brink, D.M. (1970). Phys. Lett. 32B, 149.

Bibliography

Ripka, G. and Porneuf, M. (Eds.) (1975). Nuclear Self-Consistent Fields, N. Holland, Amsterdam.

9. Finite Nuclei

In the last section, matter at subnuclear densities is being treated as a uniform gas consists of neutrons and protons which interact among each other according to the Skyrme potentials. In reality matter at such densities is frequently composed of giant nuclei which may sometimes coexist with a sea of neutrons. The presence of nuclei will necessitate the study of the role of surface energy of the nuclei. Recall that in the nuclear semi-empirical mass formula, the surface energy term contributes substantially to the total binding energy,although its contribution to the binding energy per nucleon, B/A, decreases with increasing A (see Figure 6-2). As matter density increases nuclei tends to increase in size diminishing the role of surface energy. As density is further increased, the ground state of the system will allow a sea of neutrons to exist outside the nuclei. This situation is commonly called the neutron drip, as if neutrons were dripping out of the giant nuclei. Furthermore,the treatment of the Coulomb energy is also different for the case of nuclei as it is for uniform proton distributions. The Coulomb energy problem can however be handled approximately with minor difficulty. In general, the study of matter consisting of giant nuclei is complicated, especially the presence of nuclei allows energy to be stored in nuclear excited states making the determination of the thermal properties and the equation of state of the system difficult. These problems have not been completely studied and more and more results will be coming in gradually. We shall in this section introduce the general technique of studying finite nuclei by means of nuclear potentials, in particular, the Skyrme potentials.

It is known that nuclei which possess the following numbers of neutrons or protons are particularly stable:

2, 8, 20, 50, 82, 126.

These numbers are called the magic numbers and the experimental evidence for their existence is very strong. Thus, particularly stable nuclei are $^{4}_{2}He$, $^{16}_{8}O$, $^{40}_{20}Ca$, $^{48}_{20}Ca$, $^{208}_{82}Pb$, which are doubly magic (both neutron and proton numbers correspond to magic numbers). The existence of the magic numbers can best be explained by the nuclear shell model. In this theory single-nucleon states are described by angular momentum states of an overall potential. The magic numbers correspond to closed shells of protons or neutrons in the shell model and are therefore particularly stable. The success of the shell model suggests one may profitably be considering single-nucleon states in the Hartree-Fock wave function that are angular momentum states. Such single-nucleon states may give better results than other types of states such as the harmonic oscillator states or square-well states. This consideration is relevant because the Hartree-Fock wave function is a very restricted form of the many-body wave function and for it to be a useful trial wave function it should be carefully chosen.

The single-nucleon states will be formed by combining the spin wave function χ_σ with the orbital wave functions of a central potential into total angular mementum states. The orbital wave functions are given by spherical harmonics $Y_{\ell,m}(\hat{r})$ which are the eigenfunctions of the orbital angular momentum operators, $\vec{\ell} = \vec{r} \times \vec{p}$, in the sense that,

$$\vec{\ell}^{\,2}\, Y_{\ell,m}(\hat{r}) = \hbar^2\, \ell(\ell+1)\, Y_{\ell,m}(\hat{r}) \quad ,$$

$$\ell_z\, Y_{\ell,m}(\hat{r}) = \hbar\, m\, Y_{\ell,m}(\hat{r}) \ . \qquad (9.1)$$

where ℓ and m are integers (see Appendix B.2). The total angular momentum of the state is $\vec{j} = \vec{\ell} + \vec{s}$. States of definite j are denoted by $\mathcal{Y}_{\ell jm}(r)$ which satisfy:

$$\vec{j}^{\,2}\, \mathcal{Y}_{\ell jm}(\hat{r}) = \hbar^2\, j(j+1)\, \mathcal{Y}_{\ell jm}(\hat{r}) \quad ,$$

$$j_z\, \mathcal{Y}_{\ell jm}(\hat{r}) = \hbar\, m\, \mathcal{Y}_{\ell jm}(\hat{r}) \ . \qquad (9.2)$$

For each ℓ there are two ways of forming the j states: $j = \ell + \frac{1}{2}$ and $j = \ell - \frac{1}{2}$. Because of this, the ℓ index is carried along. Explicitly,

$$\mathcal{Y}_{\ell jm} = C_+\, Y_{\ell,m-\frac{1}{2}}\chi_\uparrow + C_-\, Y_{\ell,m+\frac{1}{2}}\chi_\downarrow \qquad (\text{for } j = \ell + \tfrac{1}{2}) \ ,$$

$$\mathcal{Y}_{\ell jm} = -C_-\, Y_{\ell,m-\frac{1}{2}}\chi_\uparrow + C_+\, Y_{\ell,m+\frac{1}{2}}\chi_\downarrow \qquad (\text{for } j = \ell + \tfrac{1}{2}) \ ,$$

where

$$C_\pm = \sqrt{\frac{\ell \pm m + \frac{1}{2}}{2\ell+1}} \qquad (9.3)$$

are the Clebsch-Gordan coefficients. The single-nucleon states will be written as:

$$\psi_a(\vec{r}) = \frac{R_a(r)}{r} \mathcal{Y}_{\ell jm}(\hat{r}) \zeta_q , \qquad (9.4)$$

where $a = (n_a,\ell,j,q)$, with n_a being the principal quantum number, specifies the state.

The Hartree-Fock equations will be employed to determine ψ_a self-consistently. This, however, can lead to horrendous computational routines if the problem is not in some way simplified. One obvious simplification is to consider spherically symmetric nuclei. This will reduce a three-dimensional configuration to essentially a one-dimensional configuration, and the degree of difficulties in computation is greatly reduced. It is also true that we shall be interested mainly in such nuclei since they are usually more tightly bind. In terms of ψ_a this means all j-shells will be closed, and for every j there will be (2j+1) proton- (or neutron-) states. For spherically symmetric nuclei the matrix elements of the Hamiltonian can be reduced to depend on just the following density functions:

(1) number densities: $$n_q(r) = \sum_j |\psi_j|^2 \delta(q_j,q) , \qquad (9.5)$$

(2) kinetic energy densities: $$\tau_q(r) = \sum_j |\vec{\nabla}\psi_j|^2 \delta(q_j,q) , \qquad (9.6)$$

(3) spin densities: $$\vec{J}_q(r) = i\sum_j \psi_j^+ (\vec{\sigma}\times\vec{\nabla}\psi_j)\delta(q_j,q) , \qquad (9.7)$$

where $q = (n,p)$ with n=neutron, and p=proton. Further, define $n_B=n_n+n_p$, $\tau=\tau_n+\tau_p$, and $J=J_n+J_p$.

We shall now proceed to evaluate the matrix elements of the Hamiltonian (7.23) with v given by (8.1):(*)

(1) kinetic energy terms:

$$T = \sum_i \int d^3r\, \psi_i^+ \left(-\frac{\hbar^2}{2m}\nabla^2\right) \psi_i = \int d^3r \sum_i \frac{\hbar^2}{2m} |\vec{\nabla}\psi_i|^2 = \int d^3r \frac{\hbar^2}{2m}\tau , \qquad (9.8)$$

(2) terms proportional to t_0:

(*) *The following derivations are due to* Vautherin and Brink (1972).

$$V_0 = \frac{1}{2}t_0 \sum_{i,j} \int d^3x d^3y \psi_i^+(\vec{x})\psi_j^+(\vec{y})\delta(\vec{x}-\vec{y})(1+x_0 P_\sigma)\{\psi_i(\vec{x})\psi_j(\vec{y})-\psi_j(\vec{x})\psi_i(\vec{y})\} \ . \tag{9.9}$$

Antisymmetrization of the single-nucleon wave functions in the last paranthesis may be replaced by:

$$\{\psi_i(\vec{x})\psi_j(\vec{y})-\psi_j(\vec{x})\psi_i(\vec{y})\} = (1 - P_M P_\sigma P_\tau)\psi_i(\vec{x})\psi_j(\vec{y}) \ , \tag{9.10}$$

where the operators P_M, P_σ, P_τ are introduced to accomplish the exchange operations. Since the potential is given by a δ-function, contributions to this interaction from the two-nucleon wave functions can only come from relative S-waves which are non-zero at $(\vec{x}-\vec{y}) = 0$. Such two-nucleon states are also even under the exchange of the spatial coordinates, $\vec{x} \leftrightarrow \vec{y}$, and have eigenvalue equal to +1 uner P_M. Hence P_M in (9.10) may be replaced by 1. Making use of P_σ and P_τ given by (7.11) and (7.17), and the fact that $P_\sigma P_\sigma = 1$,

$$V_0 = \frac{1}{2} t_0 \sum_{i,j} \int d^3x d^3y \psi_i^+(\vec{x})\psi_j^+(\vec{y})\delta(\vec{x}-\vec{y})\{1-x_0 P_\tau + P_\sigma(x_0 - P_\tau)\}\psi_i(\vec{x})\psi_j(\vec{y}) . \tag{9.11}$$

The isospin matrix elements of P_τ are simply:

$$\left(\zeta_{q_1}^+(1)\zeta_{q_2}^+(2) \ P_\tau \ \zeta_{q_1}(1)\zeta_{q_2}(2)\right) = \delta(q_1,q_2) \ . \tag{9.12}$$

The matrix elements of $\vec{\sigma}_1 \cdot \vec{\sigma}_2$ will be shown below to be zero when we sum over all closed shell states. Since all states in a closed shell will have the same R_a functions, cancellations within the set of states is possible. In other words, it is claimed that for a closed shell when all (2j+1) states are included then:

$$\sum_{m=-j}^{j} \int d\Omega \ \mathcal{Y}_{\ell jm}^+(\hat{r}) \ \vec{\sigma} \ \mathcal{Y}_{\ell jm}(\hat{r}) = 0 \ , \qquad (d\Omega = d\cos\theta d\phi) \ . \tag{9.13}$$

This can be demonstrated explicitly. Let us consider the case with $j = \ell + \frac{1}{2}$; the case for $j = \ell - \frac{1}{2}$ follows in exactly the same manner. We evaluate each component of $\vec{\sigma}$ separately:

(a) $\int d\Omega \ \mathcal{Y}_{\ell jm}^+(\hat{r}) \ \sigma_x \ \mathcal{Y}_{\ell jm}(\hat{r}) = C_+ C_- \int d\Omega \left(Y_{\ell,m-\frac{1}{2}}^* Y_{\ell,m+\frac{1}{2}} + Y_{\ell,m+\frac{1}{2}}^* Y_{\ell,m-\frac{1}{2}}\right) = 0$,

The integral vanishes due to the orthogonality of the spherical harmonics.

(b) $\int d\Omega \ \mathcal{Y}^{+}_{\ell jm}(\hat{r})\ \sigma_y\ \mathcal{Y}_{\ell jm}(\hat{r}) = ic_+c_-\int d\Omega\left(Y^*_{\ell,m-\frac{1}{2}}Y_{\ell,m+\frac{1}{2}}-Y^*_{\ell,m+\frac{1}{2}}Y_{\ell,m-\frac{1}{2}}\right) = 0$,

(c) $\int d\Omega \ \mathcal{Y}^{+}_{\ell jm}(\hat{r})\ \sigma_z\ \mathcal{Y}_{\ell jm}(\hat{r}) = \int d\Omega\ \{c_+^2|Y_{\ell,m-\frac{1}{2}}|^2 - c_-^2|Y_{\ell,m+\frac{1}{2}}|^2\}$,

(d) $\int d\Omega \ \mathcal{Y}^{+}_{\ell j-m}(\hat{r})\ \sigma_z\ \mathcal{Y}_{\ell j-m}(\hat{r}) = \int d\Omega\ \{c_-^2|Y_{\ell,-m-\frac{1}{2}}|^2 - c_+^2|Y_{\ell,-m+\frac{1}{2}}|^2\}$.

Since $|Y_{\ell,m}|^2 = |Y_{\ell,-m}|^2$, (c)+(d) = 0. Hence, the spin matrix elements sum up to zero as soon as both $\pm m$ states belonging to the same j are present. This is a stronger result than (9.13) which is all we need for the present. Making use of (9.12) and (9.13), we find:

$$V_0 = \frac{1}{2}\,t_0\sum_{i,j}\int d^3x d^3y \psi_i^+(\vec{x})\psi_j^+(\vec{y})\delta(\vec{x}-\vec{y})\{1-x_0\delta(q_i,q_j)+\frac{1}{2}\left(x_0-\delta(q_i,q_j)\right)\}\psi_i(\vec{x})\psi_j(\vec{y})$$

$$= \frac{1}{2}\,t_0\int d^3r\ \{n_B(r)^2(1+\frac{1}{2}x_0)-\left(n_n(r)^2+n_p(r)^2\right)(x_0+\frac{1}{2})\}\ . \qquad (9.14)$$

(3) terms proportional to t_1:

$$V_1 = -\frac{1}{8}\,t_1\sum_{i,j}\int d^3x d^3y \psi_i^+(\vec{x})\psi_j^+(\vec{y})\delta(\vec{x}-\vec{y})\left(\nabla_x^2+\nabla_y^2-2\vec{\nabla}_x\cdot\vec{\nabla}_y\right)(1-P_M P_\sigma P_\tau)\psi_i(\vec{x})\psi_j(\vec{y})$$

$$+ \text{H.c.} \qquad (9.15)$$

where H.c. denotes the Hermitian conjugate of the first term and contains the second part of the interaction term which is proportional to t_1. Again, P_M may be replaced by +1 since $\vec{k}^2$ is a second order differential operator in the relative coordinates and will not change the parity of the state,and V_1 may be written as:

$$V_1 = -\frac{1}{8}\,t_1\sum_{i,j}\int d^3x d^3y \psi_i^+(\vec{x})\psi_j^+(\vec{y})\delta(\vec{x}-\vec{y})\left(\nabla_x^2+\nabla_y^2-2\vec{\nabla}_x\cdot\vec{\nabla}_y\right)\left(1-\frac{1}{2}(1+\vec{\sigma}_1\cdot\vec{\sigma}_2)P_\tau\right)\psi_i(\vec{x})\psi_j$$

$$+ \text{H.c.} \qquad (9.16)$$

The same derivations for (9.13) may be used to show that:

$$\sum_{m=-j}^{j}\int d\Omega\ \mathcal{Y}_{\ell jm}(\hat{r})\ \vec{\sigma}\ \nabla^2\ \mathcal{Y}_{\ell jm}(\hat{r}) = 0 \ , \qquad (9.17)$$

since $\nabla^2\mathcal{Y}_{\ell jm} = -\ell(\ell+1)r^{-2}\mathcal{Y}_{\ell jm}$. Thus,

$$V_1 = -\frac{1}{8} t_1 \sum_{i,j} \int d^3x d^3y \psi_i^+(\vec{x}) \psi_j^+(\vec{y}) \delta(\vec{x}-\vec{y}) \left(\nabla_x^2+\nabla_y^2-2\vec{\nabla}_x\cdot\vec{\nabla}_y\right)\left(1-\frac{1}{2}\delta(q_i,q_j)\right)\psi_i(\vec{x})\psi_j(\vec{y}) +$$

$$-\frac{1}{8} t_1 \sum_{i,j} \int d^3x d^3y \psi_i^+(\vec{x}) \psi_j^+(\vec{y}) \delta(\vec{x}-\vec{y}) \left(\vec{\nabla}_x\cdot\vec{\nabla}_y\right)\left(\vec{\sigma}_1\cdot\vec{\sigma}_2\right)\delta(q_i,q_j)\psi_i(\vec{x})\psi_j(\vec{y}) + \text{H.c.} \tag{9.18}$$

The evaluation of the term containing $\left(\vec{\nabla}_x\cdot\vec{\nabla}_y\right)\left(\vec{\sigma}_1\cdot\vec{\sigma}_2\right)$ requires some doing. Its contribution is minor and will be dropped in this calculation. For details of its evaluation see Vautherin and Brink (1972).

Let us next show the following simplification:

$$\sum_i \psi_i^+ \vec{\nabla} \psi_i = \frac{1}{2}\vec{\nabla}(n_n+n_p) , \tag{9.19}$$

Because $\vec{\nabla}\left(\sum_i \psi_i^+ \psi_i\right) = \sum_i \{\psi_i^+\vec{\nabla}\psi_i + (\vec{\nabla}\psi_i^+)\psi_i\}$, to establish (9.19) we only need to show that the second term on the right is equal to the first, or what is the same:

$$\sum_i \psi_i^T \nabla \psi_i^* = \sum_i \psi_i^+ \nabla \psi_i , \tag{9.20}$$

where T denotes the transpose of the spin wave function. For ψ_i given by (9.4) we have explicitly: (for $j=\ell+\frac{1}{2}$)

$$\mathcal{Y}_{\ell jm} = C_+ Y_{\ell,m-\frac{1}{2}} \chi_\uparrow + C_- Y_{\ell,m+\frac{1}{2}} \chi_\downarrow ,$$

and

$$\begin{aligned}\mathcal{Y}_{\ell j-m} &= C_- Y_{\ell,-m-\frac{1}{2}}\chi_\uparrow + C_+ Y_{\ell,-m+\frac{1}{2}}\chi_\downarrow \\ &= (-1)^{m+\frac{1}{2}}\{ C_- Y^*_{\ell,m+\frac{1}{2}}\chi_\uparrow - C_+ Y^*_{\ell,m-\frac{1}{2}}\chi_\downarrow \} \\ &= (-1)^{m+\frac{1}{2}} (i\sigma_y)\{ C_- Y^*_{\ell,m+\frac{1}{2}}\chi_\downarrow + C_+ Y^*_{\ell,m-\frac{1}{2}}\chi_\uparrow \} \\ &= (-1)^{m+\frac{1}{2}} (i\sigma_y) \mathcal{Y}^*_{\ell jm} ,\end{aligned} \tag{9.21}$$

where we make use of: $Y_{\ell,-m} = (-1)^m Y^*_{\ell m}$. Therefore, for any operator $\mathcal{O}$ which commutes with σ_y, $[\mathcal{O}, \sigma_y] = 0$:

$$\begin{aligned}\mathcal{Y}^T_{\ell j-m}\, \mathcal{O}\, \mathcal{Y}^*_{\ell j-m} &= (-1)^{m+\frac{1}{2}} \mathcal{Y}^+_{\ell jm}(-i\sigma_y)\, \mathcal{O}\, (i\sigma_y)(-1)^{m+\frac{1}{2}} \mathcal{Y}_{\ell jm} \\ &= \mathcal{Y}^+_{\ell jm}\, \mathcal{O}\, \mathcal{Y}_{\ell jm} .\end{aligned} \tag{9.22}$$

This shows the equality (9.20) since both $\pm m$ states are included in the summation. With (9.19) it follows that:

$$\sum_{i,j} \int d^3x d^3y \{\psi_i^+(\vec{x})\nabla_x^2\psi_i(\vec{x})\} \{\psi_j^+(\vec{y})\psi_j(\vec{y})\}\delta(\vec{x}-\vec{y}) = \sum_i \int d^3r \{\psi_i^+\nabla^2\psi_i\} n_B(r)$$

$$= \int d^3r \{n_B(r) \sum_i \left(\vec{\nabla}(\psi_i^+\overleftrightarrow{\nabla}\psi_i) - \vec{\nabla}\psi_i^+\cdot\vec{\nabla}\psi_i\right) - (\psi_i^+\overleftrightarrow{\nabla}\psi_i)\cdot\vec{\nabla}n_B\}$$

$$= \int d^3r \{n_B(r)\left(\frac{1}{2}\vec{\nabla}(\vec{\nabla}n_B) - \tau\right) - \frac{1}{2}\vec{\nabla}n_B\cdot\vec{\nabla}n_B\}$$

$$= \int d^3r \{n_B(r)\left(\frac{1}{2}\nabla^2 n_B - \tau + \frac{1}{2}\nabla^2 n_B\right\} .$$

Also,

$$\sum_{i,j} \int d^3x d^3y \ (\psi_i\vec{\nabla}\psi_i)(\psi_j\vec{\nabla}\psi_j) \ \delta(\vec{x}-\vec{y})$$

$$= \int d^3r \frac{1}{4} (\vec{\nabla}n_B\cdot\vec{\nabla}n_B) = -\frac{1}{4}\int d^3r \ n_B(\nabla^2 n_B) .$$

Hence,

$$V_1 = -\frac{1}{8} t_1 \int d^3r\{ \left(2n_B\nabla^2 n_B - 2n_B\tau - \frac{1}{2}n_B\nabla^2 n_B\right) - \frac{1}{2}\left(2n_n\nabla^2 n_n - 2n_n\tau_n - \frac{1}{2}n_n\nabla^2 n_n\right) +$$

$$- \frac{1}{2}\left(2n_p\nabla^2 n_p - 2n_p\tau_p - \frac{1}{2}n_p\nabla^2 n_p\right)\} + \text{H.c.} \qquad (9.23)$$

(4) terms proportional to t_2:

$$V_2 = \frac{1}{8} t_2 \sum_{i,j} \int d^3x d^3y \delta(\vec{x}-\vec{y}) \left((\vec{\nabla}_x - \vec{\nabla}_y)\psi_i^+(\vec{x})\psi_j^+(\vec{y})\right)\cdot\left((\vec{\nabla}_x-\vec{\nabla}_y)(1-P_M P_\sigma P_\tau)\psi_i(\vec{x})\psi_j(\vec{y})\right) .$$

$$(9.24)$$

In this case the two-nucleon states contributing to the matrix elements must be in relative P-states due to the gradient operator. Hence, P_M has to be replaced by -1 in the matrix elements above, which results in the following replacements:

$$(1-P_M P_\sigma P_\tau) \rightarrow (1+P_\sigma P_\tau) \rightarrow \left(1+\frac{1}{2}(1+\vec{\sigma}_1\cdot\vec{\sigma}_2)\delta(q_i,q_j)\right) \rightarrow 1 + \frac{1}{2}\delta(q_i,q_j) ,$$

$$(9.25)$$

where the last line is due to the use of (9.17) and by dropping the $(\vec{\nabla}_x\cdot\vec{\nabla}_y)(\vec{\sigma}_1\cdot\vec{\sigma}_2)$ part. Further reductions give:

$$V_2 = \frac{1}{8} t_2 \sum_{i,j}\int d^3x d^3y \delta(\vec{x}-\vec{y}) \left((\vec{\nabla}_x-\vec{\nabla}_y)\psi_i^+(\vec{x})\psi_j^+(\vec{y})\right)\cdot\left((\vec{\nabla}_x-\vec{\nabla}_y)\psi_i(\vec{x})\psi_j(\vec{y})\right)\left(1+\frac{1}{2}\delta(q_i,q_j)\right)$$

$$= \frac{1}{8} t_2 \sum_{i,j}\int d^3r\{2(\vec{\nabla}\psi_i^+\vec{\nabla}\psi_i)(\psi_j^+\psi_j) - 2(\psi_i^+\vec{\nabla}\psi_i)\cdot(\vec{\nabla}\psi_j^+\psi_j)\}\left(1+\frac{1}{2}\delta(q_i,q_j)\right)$$

$$= \frac{1}{8} t_2 \int d^3r \ \{2n_B\tau - \frac{1}{2}(\vec{\nabla}n_B)^2 + n_n\tau_n - \frac{1}{4}(\vec{\nabla}n_n)^2 + n_p\tau_p - \frac{1}{4}(\vec{\nabla}n_p)^2\} .$$

Finally,

$$V_2 = \frac{1}{16} t_2 \int d^3r \{ 4n_B\tau + 2n_n\tau_n + 2n_p\tau_p + 2n_B\nabla^2 n_B + \tfrac{1}{2} n_n\nabla^2 n_n + \tfrac{1}{2} n_p\nabla^2 n_p \} \ . \quad (9.26)$$

(5) the three-body term is given by:

$$V_3 = \frac{1}{6} t_3 \sum_{i,j,k} \int d^3x_1 d^3x_2 d^3x_3 \delta(\vec{x}_1-\vec{x}_2)\delta(\vec{x}_2-\vec{x}_3)\psi_i^+(\vec{x}_1)\psi_j^+(\vec{x}_2)\psi_k^+(\vec{x}_3) \cdot$$

$$\cdot\{ 1 - P_M(12)P_\sigma(12)P_\tau(12) - P_M(23)P_\sigma(23)P_\tau(23) - P_M(13)P_\sigma(13)P_\tau(13) +$$

$$+ P_M(12)P_\sigma(12)P_\tau(12)P_M(23)P_\sigma(23)P_\tau(23) +$$

$$+ P_M(13)P_\sigma(13)P_\tau(13)P_M(23)P_\sigma(23)P_\tau(23) \}\psi_i(\vec{x}_1)\psi_j(\vec{x}_2)\psi_k(\vec{x}_3) \ . \quad (9.27)$$

Due to the two δ-functions in the three-body potential, the Majorana operator P_M acting on any pair of particles can be replaced by +1, and the vanishing of the $\vec{\sigma}$ matrix elements permits the replacement of $P_\sigma \to \frac{1}{2}$. Thus,

$$V_3 = \frac{1}{6} t_3 \sum_{i,j,k} \int d^3x_1 d^3x_2 d^3x_3 \delta(\vec{x}_1-\vec{x}_2)\delta(\vec{x}_2-\vec{x}_3)\psi_i^+(\vec{x}_1)\psi_j^+(\vec{x}_2)\psi_k^+(\vec{x}_3) \cdot$$

$$\cdot\{ 1 - \tfrac{1}{2}\delta(q_i,q_j) - \tfrac{1}{2}\delta(q_j,q_k) - \tfrac{1}{2}\delta(q_i,q_k) + \tfrac{1}{4}\delta(q_i,q_j)\delta(q_i,q_k) +$$

$$+ \tfrac{1}{4}\delta(q_i,q_k)\delta(q_j,q_k)\} \ \psi_i(\vec{x}_1)\psi_j(\vec{x}_2)\psi_k(\vec{x}_3)$$

$$= \frac{1}{6} t_3 \int d^3r \{ n_B^3 - \tfrac{3}{2} n_B(n_n^2+n_p^2) + \tfrac{1}{2}(n_n^3+n_p^3)\}$$

$$= \frac{1}{4} t_3 \int d^3r \ n_B n_n n_p \ . \quad (9.28)$$

(6) the spin-orbit term:

$$V_{LS} = \frac{i}{8} W_0 \sum_{i,j} \int d^3x d^3y \psi_i^+(\vec{x})\psi_j^+(\vec{y}) (\vec{\sigma}_1+\vec{\sigma}_2)\cdot\big((\overleftarrow{\nabla}_x-\overleftarrow{\nabla}_y)\times\delta(\vec{x}-\vec{y}) (\vec{\nabla}_x-\vec{\nabla}_y)\big)\times$$

$$\times(1-P_M P_\sigma P_\tau)\psi_i(\vec{x})\psi_j(\vec{y}) \ , \quad (9.29)$$

where $\leftarrow$ indicates that the corresponding gradient operators are acting onto the left. This force acts in the triplet-P states only, which are states of the lowest non-vanishing spin S and orbital L, and for these one can replace P_M by -1. Further, making use of the symmetry in 1 and 2 indices of the operators, V_{LS} reduces to:

$$V_{LS} = \frac{i}{4} W_0 \sum_{i,j} \int d^3x_1 d^3x_2 \psi_i^+(\vec{x}_1)\psi_j^+(\vec{x}_2) \{\overleftarrow{\nabla}_1 \cdot \delta(\vec{x}_1-\vec{x}_2)(\vec{\nabla}_1\times\vec{\sigma}_1) - (\overleftarrow{\nabla}_1\times\delta(\vec{x}_1-\vec{x}_2)\vec{\nabla}_2)\cdot\vec{\sigma}_1 +$$

$$- (\overleftarrow{\nabla}_2\times\delta(\vec{x}_1-\vec{x}_2)\vec{\nabla}_1)\cdot\vec{\sigma}_1 + (\overleftarrow{\nabla}_1\times\delta(\vec{x}_1-\vec{x}_2)\vec{\nabla}_1)\cdot\vec{\sigma}_2\}\{1+\delta(q_i,q_j)\}\psi_i(\vec{x}_1)\psi_j(\vec{x}_2) \; . \tag{9.30}$$

Integration by parts performs the following reductions on each of the terms inside the brackets:

(a) $$\sum_{i,j} \int d^3x_1 d^3x_2 \psi_i^+(\vec{x}_1)\psi_j^+(\vec{x}_2)\{\overleftarrow{\nabla}_1\cdot\delta(\vec{x}_1-\vec{x}_2)(\vec{\nabla}_1\times\vec{\sigma}_1)\}\psi_i(\vec{x}_1)\psi_j(\vec{x}_2)$$

$$= -\sum_{i,j} \int d^3x_1 d^3x_2 \psi_i^+(\vec{x}_1)\{\vec{\nabla}_1\cdot\vec{\nabla}_1\times\vec{\sigma}\psi_i(\vec{x}_1)\}\delta(\vec{x}_1-\vec{x}_2)\psi_j^+(\vec{x}_2)\psi_j(\vec{x}_2)$$

$$-\sum_{i,j} \int d^3x_1 d^3x_2 \psi_i^+(\vec{x}_1)(\vec{\nabla}_1\times\vec{\sigma}_1)\psi_i(\vec{x}_1)\;\{\vec{\nabla}_1\delta(\vec{x}_1-\vec{x}_2)\}\psi_j^+(\vec{x}_2)\psi_j(\vec{x}_2)$$

$$= 0 \;-\sum_{i,j} \int d^3x_1 d^3x_2 \psi_j^+(\vec{x}_1)(\vec{\nabla}_1\times\vec{\sigma}_1)\psi_i(\vec{x}_1)\delta(\vec{x}_1-\vec{x}_2)\cdot\{\vec{\nabla}_2\{\psi_j^+(\vec{x}_2)\psi_j(\vec{x}_2)\}\}$$

$$= -i\int d^3r\; \vec{J}\cdot\vec{\nabla}n_B\,, \tag{9.31}$$

where we make use of the fact that $(\vec{\nabla}_1\cdot\vec{\nabla}_1\times\vec{\sigma}) = 0$ and $\vec{\nabla}_1\delta(\vec{x}_1-\vec{x}_2) = -\vec{\nabla}_2\delta(\vec{x}_1-\vec{x}_2)$.

(b) $$-\sum_{i,j} \int d^3x_1 d^3x_2 \psi_i^+(\vec{x}_1)\psi_j^+(\vec{x}_2)\{\overleftarrow{\nabla}_1\times\delta(\vec{x}_1-\vec{x}_2)\vec{\nabla}_2\cdot\vec{\sigma}_1\}\psi_i(\vec{x}_1)\psi_j(\vec{x}_2)$$

$$= -\int d^3x_1 d^3x_2 \{\psi_i^+(\vec{x}_1)\vec{\nabla}_1\times\vec{\sigma}\psi_i(\vec{x}_1)\}\cdot\{\psi_j^+(\vec{x}_2)\vec{\nabla}_2\psi_j(\vec{x}_2)\} +$$

$$+\int d^3x_1 d^3x_2 \{\psi_i^+(\vec{x}_1)\vec{\sigma}_1\psi_i(\vec{x}_1)\}\cdot\vec{\nabla}_1\delta(\vec{x}_1-\vec{x}_2)\times\{\psi_j^+(\vec{x}_2)\vec{\nabla}_2\psi_j(\vec{x}_2)\}$$

$$= -\frac{i}{2}\int d^3r\; \vec{J}\cdot\vec{\nabla}n_B + 0 \;, \tag{9.32}$$

where the second term of the last line vanishes because of (9.13).

(c) $$-\sum_{i,j} \int d^3x_1 d^3x_2 \psi_i^+(\vec{x}_1)\psi_j^+(\vec{x}_2)\{\overleftarrow{\nabla}_2\times\delta(\vec{x}_1-\vec{x}_2)\vec{\nabla}_1\cdot\vec{\sigma}_1\}\psi_1(\vec{x}_1)\psi_2(\vec{x}_2)$$

$$= -\sum_{i,j} \int d^3x_1 d^3x_2 \delta(\vec{x}_1-\vec{x}_2)\{(\vec{\nabla}_2\psi_j^+)\psi_j\}\cdot\{\psi_i^+\overleftrightarrow{\nabla}_1\times\vec{\sigma}_1\psi_i\}$$

$$= \frac{i}{2}\int d^3r\; \vec{J}\cdot\vec{\nabla}n_B. \tag{9.33}$$

(d) $$\sum_{i,j} \int d^3x_1 d^3x_2 \psi_i^+(\vec{x}_1)\psi_j^+(\vec{x}_2)\{\overleftarrow{\nabla}_1\times\delta(\vec{x}_1-\vec{x}_2)\vec{\nabla}_1\cdot\vec{\sigma}_2\}\psi_i(\vec{x}_1)\psi_j(\vec{x}_2)$$

$$= \sum_{i,j} \int d^3x_1 d^3x_2 \delta(\vec{x}_1-\vec{x}_2)(\vec{\nabla}_1\psi_i^+)\times(\vec{\nabla}_1\psi_i)\cdot\{\psi_j^+\overleftrightarrow{\sigma}\psi_j\} = 0 \;. \tag{9.34}$$

Hence,

$$V_{LS} = \frac{1}{2}W_0 \int d^3r \, \{\vec{\nabla}n_B\cdot\vec{J} + \vec{\nabla}n_n\cdot\vec{J}_n + \vec{\nabla}n_p\cdot\vec{J}_p\}$$

$$= -\frac{1}{2}W_0 \int d^3r \, \{n_B(\vec{\nabla}\cdot\vec{J}) + n_n(\vec{\nabla}\cdot\vec{J}_n) + n_p(\vec{\nabla}\cdot\vec{J}_p)\} \quad . \tag{9.35}$$

The matrix elements of the Hamiltonian is finally given by:

$$< H > \; = \int d^3r \, H(\vec{r})$$

$$= \int d^3r \, \{\frac{\hbar^2}{2m}\tau + (V_0+V_1+V_2+V_3+V_{LS}) + n_pV_C\} \, , \tag{9.36}$$

to which we have added the Coulomb interaction V_C:

$$V_C(\vec{r}) = \int d^3r' \, \frac{e^2 n_p(\vec{r}')}{|\vec{r}'-\vec{r}|} + V_{C,ex}(\vec{r}) \quad , \tag{9.37}$$

where the first term is just the classical Coulomb energy, to which one must add the exchange term due to the antisymmetrization of the proton wave functions. Since the Coulomb potential is not a zero-range potential, an exact treatment of the Coulomb exchange term would have destroyed the simplicity of the present approach with the Skyrme potentials. To preserve the simplicity of the Skyrme potentials, one may adopt the so-called Slater approximation (Slater 1951) for the exchange part of the Coulomb energy by making use of (3.36):

$$V_{C,ex}(\vec{r}) = -\frac{3e^2}{4}(3/\pi)^{1/3} \, n_p^{4/3} \quad . \tag{9.38}$$

It has been shown by Titin-Schnaider and Quentin (1974b) that by comparing with exact calculations this formula underestimates the Coulomb exchange part by less than 10%. With all the terms accounted for, the Hamiltonian density H(r) in the present spherically symmetric case depends only on the radial distance r and is given by:

$$H(r) = \frac{\hbar^2}{2m}\tau + \frac{1}{2}t_0\Big((1+\frac{1}{2}x_0)n_B^2 - (\frac{1}{2}+x_0)(n_n^2+n_p^2)\Big) + \frac{1}{4}(t_1+t_2)n_B\tau +$$
$$+ \frac{1}{8}(t_2-t_1)(n_n\tau_n+n_p\tau_p) + \frac{1}{16}(t_2-3t_1)n_B\nabla^2n_B + \frac{1}{32}(3t_1+t_2)(n_n\nabla^2n_n+n_p\nabla^2n_p) +$$
$$+ \frac{1}{4}t_3 \, n_B n_n n_p - \frac{1}{2}W_0(n_B\vec{\nabla}\cdot\vec{J}+n_n\vec{\nabla}\cdot\vec{J}_n+n_p\vec{\nabla}\cdot\vec{J}_p) + n_pV_C \quad . \tag{9.39}$$

(9.39) reduces to (8.13) which is the expression for the energy per particle in nuclear matter by letting $\vec{\nabla}n_B = \vec{\nabla}\cdot\vec{J} = 0$, $n_n = n_p = n_B/2$, $\tau_n = \tau_p = \tau/2$, and dropping the Coulomb energy term.

The Hartree-Fock equations for the Skyrme interactions are obtained by variations of the matrix elements of the Hamiltonian expectation value <H> induced by variations in the single-nucleon wave function, $\psi_i^+ \to \psi_i^+ + \delta\psi_i^+$, in the manner described in Section 1. The variations in the density functions are:

$$\begin{aligned} \delta n_B &= \delta\psi_i^+ \psi_i , \\ \delta\tau &= (\vec{\nabla}\delta\psi_i^+)\cdot\vec{\nabla}\psi_i , \\ \delta\vec{J} &= i\, \delta\psi_i^+ \vec{\sigma}\times\vec{\nabla}\psi_i . \end{aligned} \tag{9.40}$$

Using these expressions for (9.36) and integrating the τ term by parts, we find the following results: (the isospin of the ith nucleon is denoted by q)

$$\begin{aligned} \delta\langle H\rangle = \int d^3r\,\delta\psi_i^+ \{ & -\frac{\hbar^2}{2m}\nabla^2 + t_0\left((1+\tfrac{1}{2}x_0)n_B - (\tfrac{1}{2}+x_0)n_q\right) + \\ & + \tfrac{1}{4}(t_1+t_2)(\tau - \vec{\nabla}\cdot n_B\vec{\nabla}) + \tfrac{1}{8}(t_2-t_1)(\tau_q - \vec{\nabla}\cdot n_q\vec{\nabla}) + \tfrac{1}{8}(t_2-3t_1)\nabla^2 n_B + \\ & + \tfrac{1}{16}(3t_1+t_2)\nabla^2 n_q + \tfrac{1}{4}t_3(n_B-n_q)(n_B+n_q) + \delta(q,\tfrac{1}{2})V_C + \\ & - \tfrac{1}{2}W_0\left(\vec{\nabla}\cdot\vec{J} + \vec{\nabla}\cdot\vec{J}_q + i\vec{\nabla}n_B\cdot\vec{\sigma}\times\vec{\nabla} + i\vec{\nabla}n_q\cdot\vec{\sigma}\times\vec{\nabla}\right)\}\,\psi_i \end{aligned} \tag{9.41}$$

The Hartree-Fock equations are therefore:

$$-\vec{\nabla}\cdot\left(\frac{\hbar^2}{2m_q^*}\vec{\nabla}\psi_i\right) + \left(U_q(\vec{r}) + i\vec{W}_q\cdot\vec{\sigma}\times\vec{\nabla}\right)\psi_i = \varepsilon_i^{HF}\,\psi_i ,$$

where

$$\begin{aligned} \frac{\hbar^2}{2m_q^*} &= \frac{\hbar^2}{2m} + \tfrac{1}{4}(t_1+t_2)n_B(\vec{r}) + \tfrac{1}{8}(t_2-t_1)n_q(\vec{r}) , \\ U_q &= t_0\left((1+\tfrac{1}{2}x_0)n_B - (\tfrac{1}{2}+x_0)n_q\right) - \tfrac{1}{8}(3t_1-t_2)\nabla^2 n_B + \tfrac{1}{16}(3t_1+t_2)\nabla^2 n_q + \\ & \quad + \tfrac{1}{4}(t_1+t_2)\tau + \tfrac{1}{8}(t_2-t_1)\tau_q + \tfrac{1}{4}t_3(n_B^2-n_q^2) - \tfrac{1}{2}W_0(\vec{\nabla}\cdot\vec{J}+\vec{\nabla}\cdot\vec{J}_q) + \delta(q,\tfrac{1}{2})V_C , \\ \vec{W}_q &= \tfrac{1}{2}W_0(\vec{\nabla}n_B + \vec{\nabla}n_q) . \end{aligned} \tag{9.42}$$

For spherically symmetric nuclei these expressions can be simplified considerably. Making use of the single-nucleon states of (9.4) the density functions n_B, τ, and $\vec{J}$ given by (9.5) to (9.7) can readily be evaluated. The number density function is given by:

$$n_B = n_n + n_p = \sum_a (R_a/r)^2 \, \mathcal{Y}^+_{\ell jm} \, \mathcal{Y}_{\ell jm} \, (\zeta^+_q \zeta_q) \; . \tag{9.43}$$

The summation over closed-shell states will be demonstrated explicitly: (for $j=\ell+\frac{1}{2}$)

$$\sum_{m=-j}^{j} \mathcal{Y}^+_{\ell jm} \mathcal{Y}_{\ell jm} = \sum_{m=-j}^{j} \{ C^2_+ \, |Y_{\ell,m-\frac{1}{2}}|^2 + C^2_- \, |Y_{\ell,m+\frac{1}{2}}|^2 \}$$

$$= (2\ell+1)^{-1} \{ (2\ell+1)|Y_{\ell,\ell}|^2 + |Y_{\ell,\ell}|^2 + 2\ell |Y_{\ell,\ell-1}|^2 + 2|Y_{\ell,\ell-1}|^2 +$$

$$+ \; . \; . \; . \; . + 2|Y_{\ell,-\ell+1}|^2 + 2\ell|Y_{\ell,-\ell+1}|^2 + |Y_{\ell,-\ell}|^2 + (2\ell+1)|Y_{\ell,-\ell}|^2 \}$$

$$= \frac{(2\ell+2)}{(2\ell+1)} \sum_{m=-\ell}^{\ell} |Y_{\ell,m}|^2 = \frac{(2\ell+2)}{(2\ell+1)} \, \frac{(2\ell+1)}{4\pi} = \frac{(2j+1)}{4\pi} \; .$$

Similar results are obtained for closed-shell states belonging to $j=\ell-\frac{1}{2}$. In other words,

$$\sum_{m=-j}^{j} \mathcal{Y}^+_{\ell jm} \, \mathcal{Y}_{\ell jm} = (2j+1)/4\pi \; . \tag{9.44}$$

Hence, the number density function is given by:

$$n_B(r) = (4\pi r^2)^{-1} \sum_a (2j_a+1) \, R^2_a(r) \; , \tag{9.45}$$

where j_a refer to the specific j belonging to the set a.

To obtain the kinetic energy density $\tau(r)$, we make use of the relation:

$$\nabla^2 n_B(r) = 2 \sum_i \psi^+_i \, \nabla^2 \, \psi_i + 2\tau(r) \; , \tag{9.46}$$

so that,

$$\tau = \frac{1}{2} \nabla^2 n_B - (4\pi r)^{-1} \sum_a R_a \Big(r^{-2} \frac{d}{dr} r^2 \frac{d}{dr} \Big) (R_a/r) \; (2j_a+1) \; +$$

$$+ \; (4\pi r^4)^{-1} \sum_a (2j_a+1)\ell_a(\ell_a+1) \; R^2_a \; ,$$

where the last term on the right is due to: $\nabla^2 Y_{\ell m} = -\ell(\ell+1) Y_{\ell m}/r^2$.

Carrying out the differentiations in the radial variables and performing an integration by parts we have:

$$\tau(r) = \sum_a (2j_a+1)/4\pi\left\{\left(\frac{d(R_a/r)}{dr}\right)^2 + \frac{\ell_a(\ell_a+1)}{r^2}\left(\frac{R_a}{r}\right)^2\right\}. \qquad (9.47)$$

A direct evaluation of $\vec{J}_q(\vec{r})$, with q=n or p, requires skillful mainipulations of angular momentum recoupling schemes. Here, we shall derive the result using a simplifying argument that for a spherical symmetric system $\vec{J}_q(\vec{r})$ can only be proportional to $\hat{r}$ and not to $\hat{e}_\theta$ or $\hat{e}_\phi$, so that,

$$\begin{aligned}\vec{J}_q &= (\hat{r}\cdot\vec{J})\hat{r} = i\hat{r}\sum_i \psi_i^+ \hat{r}\cdot\vec{\sigma}\times\vec{\nabla}\psi_i = (\hat{r}/r\hbar)\sum_i \psi_i^+ \vec{\ell}\cdot\vec{\sigma}\psi_i \\ &= (\hat{r}/r\hbar^2)\sum_i \psi_i^+(\vec{j}^2-\vec{\ell}^2-\vec{s}^2)\psi_i \\ &= (\hat{r}/4\pi r)\sum_a (2j_a+1)\{j_a(j_a+1)-\ell_a(\ell_a+1)-\tfrac{3}{4}\}(R_a/r)^2. \qquad (9.48)\end{aligned}$$

Inserting these values into the Hartree-Fock equations, we find that both the average potential U and the effective mass m^* of each single-nucleon equation are spherically symmetric and the spin-orbit term reduces to the form:

$$i\vec{W}_q\cdot\vec{\sigma}\times\vec{\nabla} = W_q(i\hat{r}\cdot\vec{\sigma}\times\vec{\nabla}) = W_q(\vec{\ell}\cdot\vec{\sigma}),$$

where

$$W_q = \hat{r}\cdot\vec{W}_q = \frac{1}{2}W_0\frac{d}{dr}(n_B+n_q). \qquad (9.49)$$

Rewriting (9.42) as:

$$-\frac{\hbar^2}{2m_q^*}\nabla^2\psi_i - \left(\vec{\nabla}\frac{\hbar^2}{2m_q^*}\right)\cdot\vec{\nabla}\psi_i + \left(U_q + W_q\,\vec{\ell}\cdot\vec{\sigma}/r\right)\psi_i = \varepsilon_i^{HF}\psi_i,$$

with

$$\nabla^2 = \frac{1}{r}\frac{d^2}{dr^2}r - \frac{\vec{\ell}^2}{r^2},$$

$$\vec{\nabla}\frac{\hbar^2}{2m_q^*} = \hat{r}\frac{d}{dr}\left(\frac{\hbar^2}{2m_q^*}\right), \qquad (9.50)$$

the following equations for the radial wave functions are obtained:

$$\frac{\hbar^2}{2m_q^*}\left(-R_a'' + \frac{\ell_a(\ell_a+1)}{r^2}R_a\right) - \frac{d}{dr}\left(\frac{\hbar^2}{2m_q^*}\right)R_a' +$$

$$+ \{U_q + \frac{1}{r}\frac{d}{dr}\left(\frac{\hbar^2}{2m_q^*}\right) + \left(j_a(j_a+1)-\ell_a(\ell_a+1)-\frac{3}{4}\right)\frac{W_q}{r}\}R_a = \varepsilon_a^{HF}R_a \ , \quad (9.51)$$

where differentiations of R_a with respect to r is abbreviated by an appostrophy.

For an isolated nucleus the boundary conditions for R_a are simply that:

$$R_a = 0 \quad \text{at } r = 0 \text{ and } r = \infty \ . \qquad (9.52)$$

On the other hand, for a system of similar nuclei each of which is assumed to occupied a finite spherical volume, the boundary conditions are somewhat different and we shall come to that later.

The Hartree-Fock eqauations are then solved numerically by an iteration procedure. In these equations not only the starting wave functions are not known, but also that the Hartree-Fock energies are to be determined. To start the iteration process, first let $R_a^{(0)}$ denote the first approximations to the unknown radial wave functions. $R_a^{(0)}$ may be chosen to be the so-called Woods-Saxon wave functions which have the analytical form of the nucleus (6.3). These wave functions yield the first approximations to $n_q^{(0)}, \tau_q^{(0)}$ and $\vec{J}_q^{(0)}$ and therefore to $U_q^{(0)}$, $\vec{W}_q^{(0)}$ and $m_q^{*(0)}$. With these as inputs, an eigenvalue ε_a^{HF} can be found for a wave function satisfying the boundary conditions (9.52). The determination of the eigenvalue is done by assuming a trial ε_a^{HF}. The equation is integrated numerically from r=0. This is done by taking $R_a(0)=0$ and $R_a'(0)$ to be any convenient value since the equation is homogeneous and its exact value is only determined by the normalization condition. With $R_a(0)$ and $R_a'(0)$, the Hartree-Fock equation provides the second derivative $R_a''(0)$ to proceed to the next interval, and then to the next, and so forth. For an arbitrary ε_a^{HF}, the wave function computed will not match the boundary condition at $r = \infty$. Thus, the trial ε_a^{HF} would have to be adjusted to give the wave function the right boundary condition. Also, R_a may not have the correct number of nodes, which are specified by the principal quantum number n_a. The number of nodes in R_a between 0 and ∞ should be equal to (n_a-1). Again, ε_a^{HF} would have to be adjusted until R_a is found to possess the correct number of nodes. The first corrected wave functions $R_a^{(1)}$ and eigenvalues ε_a^{HF} are found by these criteria. $R_a^{(1)}$ are further to be normalized. With the normalization constant N_a computed from:

$$\int_0^\infty dr\,\left(R_a^{(1)}\right)^2 = N_a^2 . \tag{9.53}$$

After the first iteration, $U_q^{(1)}$, $W_q^{(1)}$, and $m_q^{*(1)}$ will then be constructed from $R_a^{(1)}$ and become inputs for a second iteration of the radial wave functions. The procedure will be repeated until the trial eigenvalues ε_a^{HF} between two consecutive iterations differ by less than a prespecified amount. Vautherin and Brink reported that about 15 to 20 iterations were needed in their calculations to achieve relative variations in eigenvalues under 10^{-4}.

Since errors in numerical operations tend to accumulate, it is important to keep the number of operations meaningfully small. A more common procedure for this problem is to replace the boundary condition at infinity by a modified boundary condition at a point reasonably close by. Let us choose a point $r=r_s$, where r_s may be several times the size of a nucleus as estimated by (6.1). In addition to the radial wave function which is obtained by integrating the differential equation outwards from r=0, we have a second radial wave function which is obtained by integrating inwards from $r=r_s$. These two solutions are then required to match at an intermediate point $r=r_M$. Denoting the solution inside of r_M by R_{aI} and that outside by R_{aO}, then the matching condition at r_M is the equality of their logarithmic derivatives:

$$\frac{R'_{aI}(r_M)}{R_{aI}(r_M)} = \frac{R'_{aO}(r_M)}{R_{aO}(r_M)} , \tag{9.54}$$

since the differential equation is homogenous. The eigenvalues are now determined by this condition which is appropriate for finite nuclei. The starting values for R_{aO} at $r=r_s$ may be chosen to be $R_{aO}(r_s)=0$ or any conveniently small value, and $R'_{aO}(r_s)$ arbitrary. The normalization constant is now computed from:

$$N_a^2 = \int_0^{r_M} dr\,\left(R_{aI}\right)^2 + \left(R_{aI}(r_M)/R_{aO}(r_M)\right)^2 \int_{r_M}^{r_s} dr\,\left(R_{aO}\right)^2 . \tag{9.55}$$

Radial wave functions which satisfy (9.54) and normalized by N_a of (9.55) will be used to construct U_q, W_q and m_q^*, from which the next iteration of R_a is to be performed.

This method gives a fairly accurate description of the particle distribution for isolated closed-shell nuclei such as O^{16}, Ca^{40}, and Pb^{208}, which have been studied by Vautherin and Brink using the SI and SII parameters.

The total energy of the system thus computed includes the kinetic energy due to the center-of-mass motion of the nucleus which must be subtracted from the total energy in order to obtain the total binding energy of the nucleus. The center-of-mass kinetic energy is just $\vec{P}^2/(2mA)$, where $\vec{P} = \sum_i \vec{p}_i$ is a sum of single-particle momenta. It splits up into two terms:

$$\frac{1}{2mA}\vec{P}^2 = \frac{1}{2mA}\sum_{i=1}^{A}\vec{p}_i^{\,2} + \frac{1}{2mA}\sum_{i\neq j}\vec{p}_i\cdot\vec{p}_j \ . \tag{9.56}$$

The first term can be taken care of easily. The second term gives rise to exchange corrections and is difficult to calculate. The second term is often neglected. In general, the center-of-mass correction with just the first term improves the final results for the binding energies.

When matter density is sufficiently high, say, a fraction of the nuclear density, the nuclei in it can no longer be treated as isolated systems. Nuclei now coexist with a sea of neutrons so that nucleon density is finite everywhere. In this case, each nucleus is assumed to occupy a spherical (Wigner-Seitz) cell with radius r_s given by (3.1). The treatment is similar to that described in the application of the Thomas-Fermi method to atomic structures. The radial wave functions now obey new boundary conditions. In a study by Bonche and Vautherin (1981), the following two types of boundary conditions at $r=r_s$ are chosen:

(I) $R_a(r_s)=0$ if ℓ_a is even and $\left.\frac{d}{dr}(R_a/r)\right|_{r=r_s} = 0$ if ℓ_a is odd, (9.57)

(II) $R_a(r_s)=0$ if ℓ_a is odd and $\left.\frac{d}{dr}(R_a/r)\right|_{r=r_s} = 0$ if ℓ_a is even. (9.58)

As explained by these authors, these two types of boundary conditions give very nearly identical numberical results in total energy. Alternative boundary conditions of requiring either $R_a(r_s) = 0$, or

$\frac{d}{dr}\left(R_a/r\right)\Big|_{r=r_s} = 0$ for all waves would produce either a hole or a lump at $r=r_s$. Both of these cases result in higher energies for the nucleus as compared with those computed with (9.57) or (9.58). Thus, (9.57) or (9.58) are chosen. The density profiles of neutron and proton distributions along the line joining two adjacent nuclei obtained by Bonche and Vautherin using the SKM parameters are shown in Figure 9-1. The proton distributions are given by the lower curves while the neutron distributions are given bv the ordinates between the upper and lower curves. The proton fraction is kept fixed at $Y_e = 0.25$. Qualitatively similar results had previously been obtained by Negele and Vautherin (1973) employing different effective interactions. From Figure 9-1 we see that as matter density increases the space between adjacent cells is being filled. When the density becomes of the order

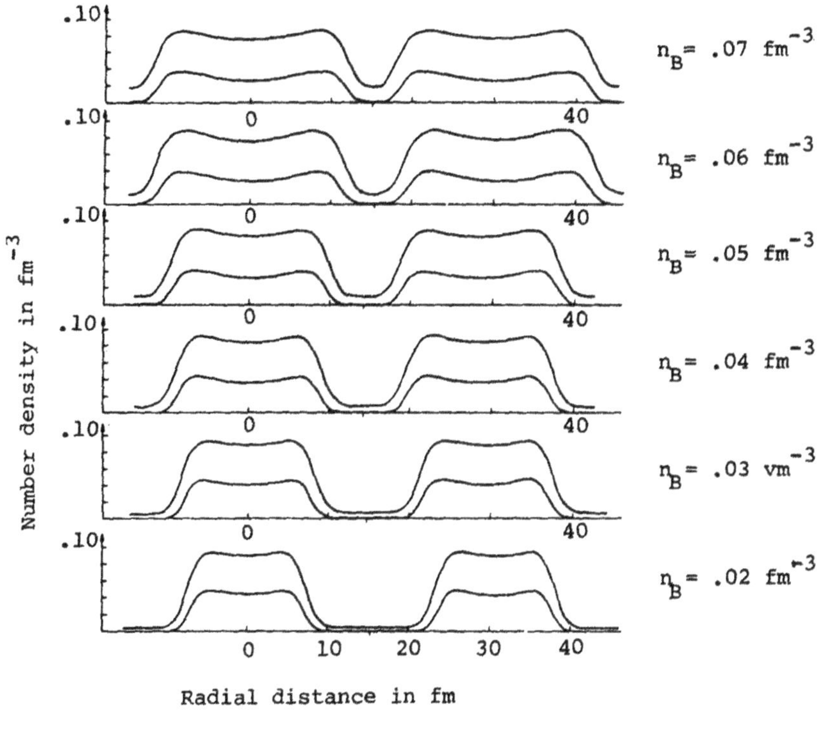

Figure 9-1. Density profiles of the neutron (upper curves) and proton (lower curves) distributions along the line joining two adjacent cells. From Bonche and Vautherin (1981).

of $n_B = 0.06\ fm^{-3}$, Bonche and Vautherin reported that the matching to the adjacent cells appears less satisfactory. This is because nuclei are nearly touching as the density is close to that of nuclear matter ($n_B = 0.12\ fm^{-3}$ for $Y_e = 0.25$). At $n_B = 0.06\ fm^{-3}$, nuclei are expected to turn into bubbles, which have smaller surface areas than giant nuclei and thus reducing the surface energy. The bubble configurations can also be obtained by the Hartree-Fock method using boundary conditions (9.57) or (9.58). The starting trial wave functions should be different, but the general method of computation is completely analogous to that discussed before.

References

Bonche, P. and Vautherin, D. (1981). Nucl. Phys. A372, 496.

Negele, J.W. and Vautherin, D. (1973). Nucl. Phys. A207, 298.

Slater, J.C. (1951). Phys. Rev. 81,385.

Titin-Schnaider, C. and Quentin, Ph. (1974a). Phys. Lett. 49B, 213.

Titin-Schnaider, C. and Quentin, Ph. (1974b). Phys. Lett. 49B, 397.

Vautherin, D. and Brink, D.M. (1970). Phys. Lett. 32B, 149.

Vautherin, D. and Brink, D.M. (1972). Phys. Rev. C5, 626.

Bibliogrpahy

De-Shalit, A. and Talmi, I. (1963). Nuclear Shell Theory, Academic Press, New York.

Mayer, M.G. and Jensen, J.H.D. (1955). Elementary Theory of Nuclear Structure, Wiley, New York.

Marmier, P. and Sheldon, E. (1970). Physics of Nuclei and Particles, Vol. 2, Academic Press, New York.

Migdal, A.B. (1967). Theory of Finite Fermi Systems and Applications to Atomic Nuclei, Wiley Interscience, New York.

Quentin, P. and Flocard, H. (1978). Ann. Rev. Nucl. Part. Sci. 28, 523.

10. Subnuclear Matter at Finite Temperatures

Topics discussed in the last few sections of Chapter 2 are limited to situations where thermal effects play a negligible role and therefore the state of matter is treated as if it is at $T = 0$. We shall now extend these topics to include thermal effects. The temperature that we shall consider is much higher than ordinary stellar temperatures and is only reached when a stellar body is going through gravitational collapse. On the other hand, the thermal energy $k_B T$ at such temperatures is still small compared with the nuclear interaction energy so that the basic nucleon interactions are not modified by the temperature. The following study will be limited to situations of this type.

We shall be mainly interested in matter densities near nuclear density. For simplicity, the system is considered to consist of uniformly distributed nuclei of a single specis. The thermal Hartree-Fock method introduced in Section 4 will be employed. Nuclear interactions are taken to be the Skyrme type. The thermal Hartree-Fock equations would have exactly the same form as (9.42) except that m_q^*, U_q and $\vec{W}_q$ are temperature dependent.

We shall first consider a simplified case in which nucleons do not conglomerate into nuclei but distribute uniformly. This would be the case if matter density is at nuclear density. In this case the Hartree-Fock single-nucleon wave functions will be plane waves, and the Hartree-Fock single-nucleon energies are:

$$\varepsilon_i^{HF} \equiv \varepsilon_{k,q} = \frac{\hbar^2}{2m_q^*} k_q^2 + V_q , \tag{10.1}$$

where $q = n$ or p, m_q^* are given by (9.42) and V_q derivable from U_q of (9.42) by setting $\vec{\nabla} n_q = 0$. In summary,

$$\frac{\hbar^2}{2m_q^*} = \frac{\hbar^2}{2m} + \frac{1}{4}(t_1+t_2)n_B + \frac{1}{8}(t_2 - t_1)n_q , \tag{10.2}$$

and,

$$V_q = t_0 \{ (1 + \frac{1}{2}x_0)n_B - (\frac{1}{2} + x_0)n_q \} + \frac{1}{4}(t_1+t_2)\tau + \frac{1}{8}(t_2-t_1)\tau_q + \frac{1}{4}t_3(n_B^2 - n_q^2) , \tag{10.3}$$

where $\tau = \tau_n + \tau_p$ and $n_B = n_n + n_p$. The density parameters are given by:

$$n_q = \pi^{-2} \int_0^\infty k^2 dk\, \bar{n}_q(k) \ , \tag{10.4}$$

$$\tau_q = \pi^{-2} \int_0^\infty k^4 dk\, \bar{n}_q(k) \ , \tag{10.5}$$

where,

$$\bar{n}_q(k) = \{ 1 + \exp\big(\beta(\varepsilon_{k,q} - \mu_q)\big)\}^{-1} \ , \tag{10.6}$$

as in (4.43). n_q and τ_q can be expressed as Fermi integrals:

$$n_q = \frac{1}{2\pi^2}\left(\frac{2m_q^*}{\hbar^2\beta}\right)^{3/2} F_{1/2}(y_q) \ , \tag{10.7}$$

$$\tau_q = \frac{1}{2\pi^2}\left(\frac{2m_q^*}{\hbar^2\beta}\right)^{5/2} F_{3/2}(y_q) \ , \tag{10.8}$$

where,

$$y_q = \beta(\mu_q - V_q) \ . \tag{10.9}$$

The most direct method to evaluate these parameters is to start out with a pair of neutron and proton densities, n_n and n_p. m_q^* are thus determined and (10.7) can be inverted to yield y_q. τ_q and V_q are now computed, and from (10.9) the chemical potentials of the nucleons.

The entropy densities of the nucleons are to be evaluated from (4.37), which in the present case is given by:

$$S_q/(k_B\Omega) = \beta \{\frac{5}{3}\frac{\hbar^2}{2m_q^*}\tau_q + n_q(V_q - \mu_q)\} \ . \tag{10.10}$$

The pressure P of the system is given by (4.23):

$$P = \beta^{-1}(S_n + S_p)/(k_B\Omega) + \mu_n n_n + \mu_p n_p - E/\Omega \ , \tag{10.11}$$

where E/Ω is the energy density of the system and is to be evaluated from (9.39) by setting all gradient terms to zero, or from (8.24), but note that these two expressions are based on different forms of the Skyrme interactions in the density dependent or the three-body term. Thus, the equation of state for such a uniform system of nucleons at finite temperatures can be constructed.

At subnuclear densities the nucleons are not distributed uniformly, but instead part of the nucleons form nuclei which is at the nuclear saturation density, while the remaining nucleons distribute in the surrounding space at a much lower density. The nucleons are therefore separated into two phases. Those which conglomerate into nuclei are in a denser phase (phase 1), while the others remaining outside are in a less dense phase (phase 2). Matter in phase 2 is also referred to as dripped matter. We shall next ignore the surface effects of the nuclei and study the coexistence of these two phases as bulk matter. The criteria for coexistence are (1) the equality of the chemical potentials of both nucleon components:

$$\mu_{n1} = \mu_{n2} \;, \quad \mu_{p1} = \mu_{p2} \;, \tag{10.12}$$

and (2) the equality of the pressures of the two phases:

$$P_1 = P_2 \;. \tag{10.13}$$

In order to see how these criteria are to be satisfied, we follow here the discussion presented by Lattimer and Ravenhall (1978). They summarized the results on uniform distributions given by (10.1) to (10.11) into a series of isobars on plots of chemical potentials versus proton concentrations (μ_q vs. x), one plot for each temperature. The proton concentration in each phase will be denoted by $x_i = n_{pi}/(n_{ni}+n_{pi})$ where i refers to phase 1 and 2. The proton concentration x used here refers to a uniform system and is the same as the proton fraction Y_p used in previous discussions. Two such plots are shown in Figure 10-1, one at $k_BT = 0$ and the other at $k_BT = 10$ MeV. The solid line corresponds to an isobar of a definite pressure. In this case, the pressure of the isobar shown is 0.115 MeV/fm^3 or 1.84×10^{32} dy/cm^2. In these plots, the neutron or proton densities are not displayed. The portion of the curve above its value at x = 0.5 will be called the upper curve, and the curve below the lower curve. The values of the chemical potential of the upper curve represent μ_n and those of the lower curve μ_p. A point in the plot will be labelled (μ_q, x).

At a given pressure P, the coexistence between dripped neutrons in phase 2 with a neutron-proton mixture in phase 1 is possible if a point $(\mu_n, 0)$ on the isobar P can be connected to a point (μ_n, x) on the same isobar by a horizontal line. This is indicated by the horizontal

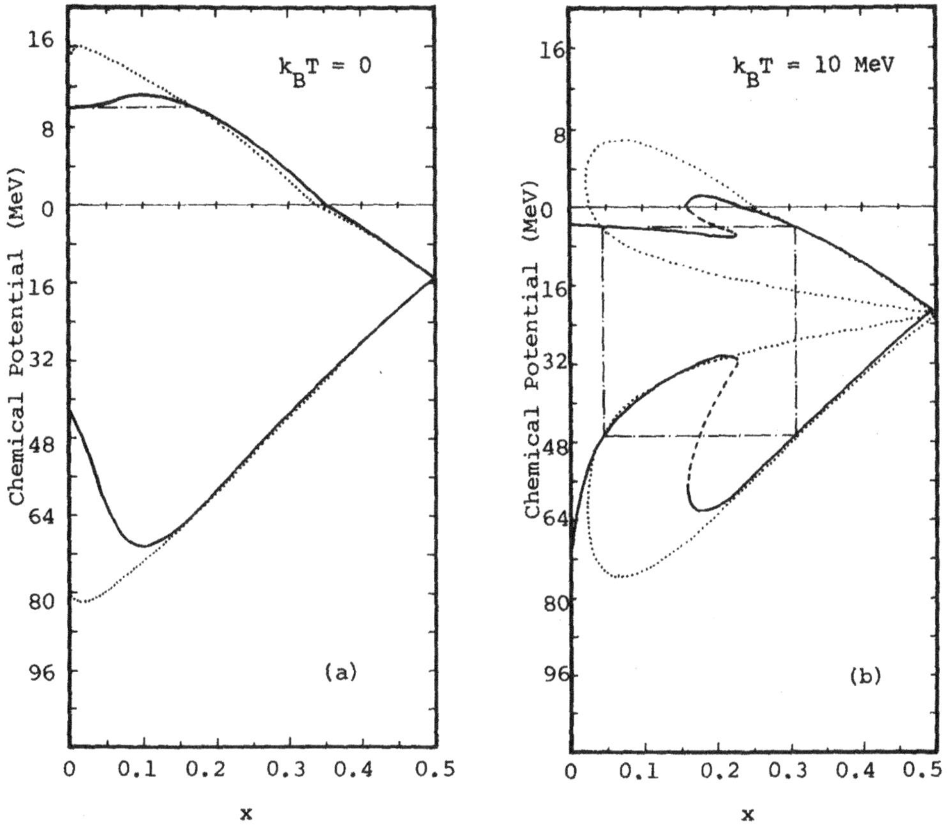

Figure 10-1. Neutron and proton chemical potentials plotted as a function of the proton concentration x along an isobar (solid line). The dashed portion has negative compressibility and is unstable. The upper curve represents μ_n and the lower curve μ_p. The dotted line is the coexistence curve.

dot-dashed lines in Figure 10-1(a). The locus of (μ_n, x) for which coexistence is possible is drawn as dotted lines (the top dotted lines). For every point on this locus there is a corresponding (μ_p, x) on the same isobar. The locus of these points are indicated by the bottom dotted lines.

We must consider also the possibility of dripped protons. In this case a neutron-proton mixture in phase 2 coexists with that in phase 1. We shall label the proton concentrations in phases 1 and 2 by subscripts 1 and 2, respectively. The possibility of coexistence is available if the pair of points (μ_{n1}, x_1) and (μ_{n2}, x_2), as well as (μ_{p1}, x_1) and (μ_{p2}, x_2), along the same isobar can be connected simultaneously by horizontal lines. In other words, these four points form the corners of a rectangle as shown in Figure 10-1(b). The number densities of each specis in each phase are to be deduced from their chemical potentials.

To express these results in terms of a nucleon density n_B and proton fraction Y_p of the system, let us introduce a quantity u, which is the fraction of space occupied by phase 1. The nucleon number density of the system is then given by:

$$n_B = (n_{n1} + n_{p1})u + (n_{n2} + n_{p2})(1 - u) , \tag{10.14}$$

and the proton fraction of the system is:

$$Y_p = n_B^{-1} \{n_{p1}u + n_{p2}(1 - u)\} , \tag{10.15}$$

u is then determined by minimizing the Helmholtz free energy F = E - TS of the system. The mass density of the system is given by:

$$\rho = m_n n_n + m_p n_p + E/\Omega c^2 , \tag{10.16}$$

where $n_n = n_{n1}+n_{n2}$ and $n_p = n_{p1}+n_{p2}$.

The electron fraction of the system is going to be equal to the proton fraction, $Y_e = Y_p$, in order to satisfy charge neutrality. The electrons will added to the entropy and pressure of the system. Chemical equilibrium of the system imposes the following condition:

$$(\mu_n + m_n c^2) = (\mu_p + m_p c^2) + \mu_e , \tag{10.17}$$

which is the same as (5.9). In the present case the neutron and proton chemical potentials are defined by (10.9) which do not include the rest mass of the nucleons, whereas μ_e is given by the relativistic expression. Under special conditions, as in the later stages of stellar collapse, neutrinos are trapped within the stellar core and the chemical potential of the neutrinos will modify the equilibrium condition (10.17). Also, at high temperature, positrons will be thermally excited. We shall neglect these components for now. Also neglected are the photon contributions to entropy and pressure.

In phase 2 where the density is low we expect the interaction energy to be negligible, and if the chemical potential of the nucleon specis is zero or negative, its number density would also vanish. In view of Figure 10-1, we do not expect to encounter dripped protons in phase 2, since the proton chemical potentials are almost always negative. We may ignore the mentioning of dripped protons except possibly at high temperatures. Also, if we follow the upper dotted lines in Figure 10-1(a), we see that it crosses the $\mu_n = 0$ line at $x = 0.34$. This implies that no dripped neutron should be pressent. Dripped neutrons in phase 2 are present at $T = 0$ only if $x_1 < 0.34$, and dripped neutron density increases with lower x_1.

The two phases will be treated as uniform systems separately. Their Helmholtz free energies $F_i = E_i - T\,S_i$ will be computed according to the chemical potentials of the components. The free energy of the total system:

$$F = uF_1 + (1 - u)F_2 - Y_p n_B (m_n - m_p)c^2\Omega \,, \tag{10.18}$$

will be minimuzed with respect to u for fixed P, T and Y_p. The last term in (10.18) accounts for the mass difference of the nucleon specis which has not been included in E_i.

The results are depicted in Figure 10-2, which is taken from Lamb, Lattimer, Pethick and Ravenhall (1981). The solid lines form the boundaries of the one-phase and two-phase systems in a temperature versus density plot. Each line corresponds to a definite proton fraction Y_p. For example, at $T = 0$ which corresponds to the abscissa of the plot, a two-phase system is formed when its density is below that indicated by the solid lines at approximately nuclear matter density. Above this density matter exists as a one-phase system, for which discussions given in the early part of this section apply.

At finite temperatures the two-phase system occurs in the region under the solid lines. At densities above 10^{14} g/cm^3 temperature has minor effects on the system as to be expected. At lower densities high temperature will keep matter from forming a two-phase system. The dashed lines in Figure 10-2 are contours along which half the nucleons are in phase 1 and the other half in phase 2.

We shall next consider the nuclear surface energy. Had we carried out the Hartree-Fock calculations for finite nuclei outlined in Section 9,

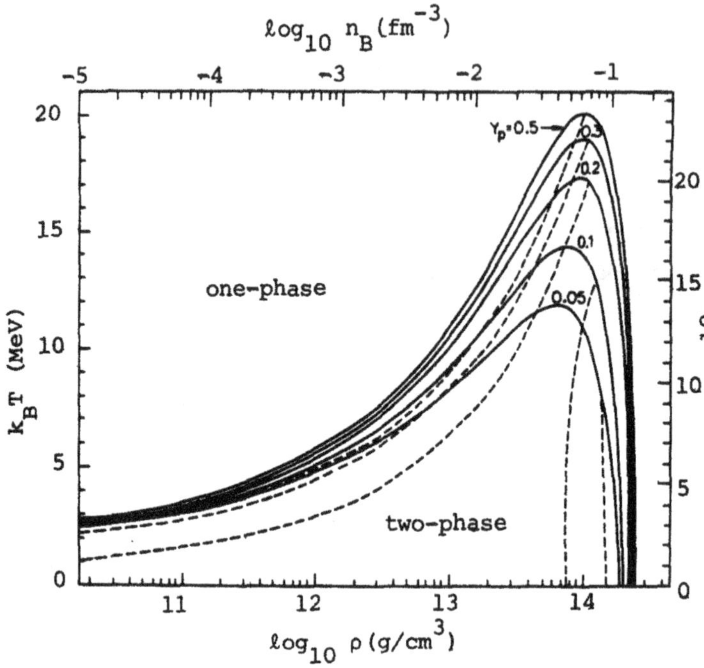

Figure 10-2. Boundaries of the two-phase region for various proton fractions Y_p are shown as solid lines The dashed lines are contours along which half of the nucleons are in phase 1. Graph reproduced from Lamb, Lattimer, Pethick and Ravenhall (1981).

the calculations can be extended to include finite temperature effects by employing the thermal Hartree-Fock equations. The thermodynamical properties of a system of finite nuclei can thus be determined from such calculations, which have been attempted by Bonche and Vautherin (1981). The task is ardous, since every specific situation of n_B, Y_p, and T requires a separate Hartree-Fock self-consistent calculation. Therefore, before any of such calculations is carried out some estimates of the region of interest should be made. We shall refer those readers who are interested in the Hartree-Fock approach to the article by Bonche and Vautherin, and we shall describe below a simplified treatment of the nuclear surface energy employing the liquid drop model.

In the following development of the liquid drop model by Baym, Bethe and Pethick (1971) each nucleus is assumed to occupy a spherical Wigner-Seitz cell of radius r_s. There will be a central high density region, referred to as the nucleus, surrounded by a low density shell. The bulk nucleon number density in the nucleus is denoted by $n_1=n_{n1}+n_{p1}$ and that exterior to the nucleus is denoted by $n_2=n_{n2}+n_{p2}$. The actual shape of the density distribution throughout the cell is denoted by n(r). To simplify discussion we shall further assume the proton concentration to be the same everywhere ($x_1 = x_2$). This is called

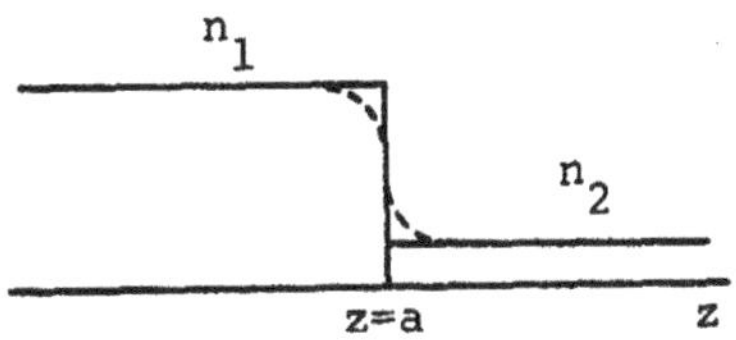

Figure 10-3. Schematic illustration of a nuclear surface between two separate phases.

the one-fluid approximation by Ravenhall, Bennett and Pethick (1972). Neglecting effects due to surface curvature, the problem is reduced to a single dimension, which we choose to be the z-axis. Imagine there is a reference plane at z = a perpendicular to the z-axis. n_1 exists to the left of the plane and n_2 to the right, as shown in Figure 10-3. n_1 and n_2 are related by conditions of bulk equilibrium (10.12) and (10.13). The location of the plane is determined by:

$$\int_{-\infty}^{a} dz\,\{n(z) - n_1\} + \int_{a}^{\infty} dz\,\{n(z) - n_2\} = 0 , \qquad (10.19)$$

where the nuclear interior is extended to $z = -\infty$ and the exterior to $z = \infty$. Since n(z) is expected to be given by n_1 or n_2 not far from z = a, such extensions should have no effects on the surface energy. Let us denote the energy per nucleon by W = E/A, which may be evaluated from (9.39) with Skyrme interactions. W is in general a function of n, $\vec{\nabla}n$, τ and $\vec{\nabla}J$ in addition to be dependent on the proton concentration x. Let us denote the dependence of W on all these variables collectively by writing W(z). Then, the surface energy per unit area of the nucleus is given by:

$$\sigma = \int_{-\infty}^{a} dz\,\{W(z)n(z) - W_1 n_1\} + \int_{a}^{\infty} dz\,\{W(z)n(z) - W_2 n_2\} , \qquad (10.20)$$

where W_1 and W_2 are the energies per nucleon at n_1 and n_2, respectively. An estimate of σ can be obtained by writing the density and energy profiles in the forms:

$$n(z) = n_2 + (n_1 - n_2)\, f(z/b) ,$$

and,

$$W(z) = W_2 + (W_1 - W_2)\, g(z/b) , \qquad (10.21)$$

where b is a surface thickness parameter, and f and g are arbitrary functions. If we further assume that $f = g$, which is to say that the energy per nucleon is a linear function of the density, then,

$$\sigma = \lambda b(W_2-W_1)(n_1-n_2) , \tag{10.22}$$

where,

$$\lambda = \int_{-\infty}^{\infty} dz\, f(z)\,\{1 - f(z)\} . \tag{10.23}$$

For a first approximation to b we may write:

$$b \simeq \eta \frac{\pi}{k_c} , \tag{10.24}$$

where η is a parameter ($\eta \simeq 1$) and k_c is determined from:

$$\frac{2}{3\pi^2} k_c^3 = (n_1 - n_2) , \tag{10.25}$$

Inserting these relations into (10.21), the approximate surface energy is found to be:

$$\sigma = \left(\frac{2\pi}{3}\right)^{1/3} \eta\, \lambda\, (W_2-W_1)(n_1-n_2)^{2/3} . \tag{10.26}$$

Multiplying σ by the area of the spherical nuclear surface, which is given by $4\pi r_A^2$, with $r_A = \{(3/4\pi)(A/n_1)\}^{1/3}$, we find the total surface energy per nucleon to be:

$$W_s = b_s(W_1-W_2)\{1 - (n_2/n_1)\}^{2/3} A^{2/3} , \tag{10.27}$$

where,

$$b_s = 2(3\pi^2)^{1/3} \lambda\eta , \tag{10.28}$$

may be treated as a phenomenological parameter. W_s in (10.27) should reduce to the surface energy term in the nuclear semi-empirical mass formula when n_2 is set equal to zero (for $x_1 = 0.5$).

In order to improve on the above treatment of the surface energy, it is necessary to make more detailed studies of the n(z) and W(z) functions in (10.20). Instead of carrying out a full-fledged Hartree-Fock calculation as described in Section 9, a Thomas-Fermi approach is usually taken (Bethe 1968). This amounts to evaluating certain Hartree-Fock quantities by means of the plane wave states instead of the Hartree-Fock single-nucleon states. For example, the following quantities of Section 9 are approximated by:

$$n_q(z) = \frac{1}{3\pi^2} k_{Fq}^3(z) ,$$

$$\tau_q(z) = \frac{1}{5\pi^2} k_{Fq}^5(z) ,$$

$$\vec{J}_q(z) = 0 . \qquad (10.29)$$

These quantities depend on z and their gradients do not vanish. The problem reduces basically to two z-dependent functions, namely, $k_{Fn}(z)$ and $k_{Fp}(z)$, or $n_n(z)$ and $n_p(z)$.

Depending on the degree of accuracy desired, the Thomas-Fermi method may be carried out by a variety of approaches. In one approach, the nucleon number density functions are first expressed by analytic forms with unknown parameters, which are then determined by a variational procedure, i.e., they are varied so as to achieve the minimum energy for the system. This approach was taken by Brueckner, Buchler, Jorna and Lombard (1968).

A second approach is to try to obtain $n_{n,p}(z)$ by an iteration procedure. $n_{n,p}(z)$ and $\vec{\nabla}n_{n,p}(z)$ must be such as to maintain constant chemical potentials $\mu_{n,p}$ throughout the cell. The chemical potentials in this case refer to the maximum Hartree-Fock energies for these components. Proton concentrations may be different in the two regions and this is referred to as the two-fluid model. Surface energy has been studied in this manner by Ravenhall, Bennett and Pethick (1972) employing the Skyrme interactions with RBP parameters (see Table 8-1). It is found to be given in the following form:

$$\sigma(x_1, T=0) \simeq 1.149 \{ 1 - 9.051(0.5-x_1)^2 \text{ MeV/fm}^2\}, \quad x_1 > 0.25$$
$$\simeq 20.5\ (x_1+0.04)^3 \text{ MeV/fm}^2, \quad x_1 < 0.25 , \qquad (10.30)$$

where the surface energy is written as a function of the proton concentration of the nucleus x_1 and temperature T. (10.30) is given by Lattimer (1981) based on calculations by Kohler (1976) and Farine, Cote and Pearson (1980).

Temperature dependence of σ may be found from the thermal Hartree-Fock formulation with roughly the same procedure described above. The temperature dependence is found to be relatively insensitive to the proton concentration in the nucleus (Lattimer 1981):

$$\frac{\sigma(x_1,T)}{\sigma(x_1,0)} = 1 - \left(\frac{T}{T_s}\right)^2 , \tag{10.31}$$

where $k_BT_s = 12.55 + 0.064(k_BT)^{1.6}$ with k_BT and k_BT_s given in units of MeV.

The system is now divided into spherical Wigner-Seitz cells of the same volume. Within each cell matter in phase 1 coexists with that in phase 2. Matter in phase 1 is further enclosed within a spherical volume of radius r_A referred to as the nucleus. The ratio of the volume occupied by the nucleus to that of the Wigner-Seitz cell is u. At some densities matter in phase 2 will occupy a spherical volume within the cell. In that case it is called a bubble. The formation of bubbles is expected to occur at densities above half of the nuclear matter density, and the bubbles should disappear completely at nuclear matter density when matter returns to a system of uniformly distributed nucleons. The departure from a uniform system would also give rise to Coulomb energy and lattice energy, and at high temperatures the lattice energy will give way to the translational energy of the nuclei which obeys a Maxwellian velocity distribution. These quantities may be estimated from a classical picture of the nucleus with various degrees of accuracy. Rather extensive discussions of them are given by Baym, Bethe and Pethick (1971) and by Lattimer (1981). The possibility for the presence of alpha particles in phase 2 should also be considered. The interested readers are referred to these articles.

The total energy of the system and thus its free energy should now include the surface energy, the Coulomb energy and translational energy. Minimization of the free energy determines the size of the nucleus within the Wigner-Seitz cell. Thus, it is possible to associate a mass number A to each nucleus and a dripped nucleon number A' to each cell. Results obtained by Lamb, Lattimer, Pethick and Ravenhall (1978) are shown in Figure 10-4 on a ρ-T plot for the case of

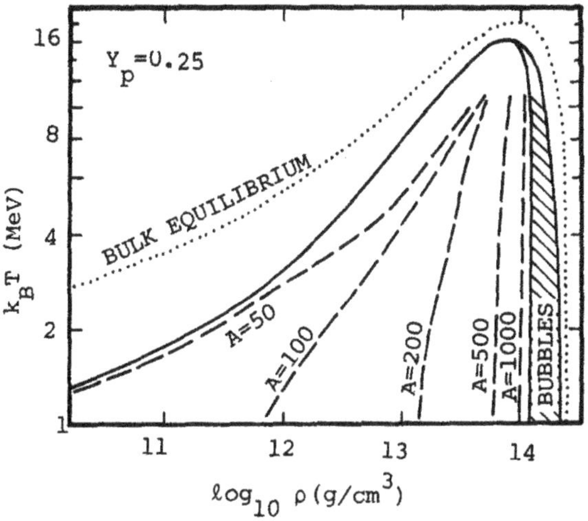

Figure 10-4. Composition of hot dense matter with Y_p=0.25 The outermost solid line shows the boundary between the one-phase and two-phase region (see text). A is the mass number of the nuclei.

$Y_p = 0.25$. The outermost solid line separates the one-phase region from the two-phase region. To be exact, it corresponds to the contour along which 10% of the nucleons in a cell conglomerates into a nucleus. This line is to be compared with the dotted line which is found from the bulk equilibrium method without the inclusion of surface, Coulomb and translational energies. Within the two-phase region the estimated mass numbers A of the nuclei are shown. A increases with density. Bubbles are formed in the hatched region of the plot.

The equation of state now relates pressure P to ρ, T and Y_p. Being a multi-dimensional relationship, it is impossible to list the equation of state numerically. In practice, the trajectory in a ρ-T-Y_p plot must be predetermined before a simple equation of state in the form of P vs. ρ can be constructed. As it has been emphasized by Bethe, Brown, Applegate and Lattimer (BBAL, 1979) that one of the most important features of the physics of supernova is the trapping of neutrinos at high matter densities. At a density of $\rho = 10^{12}$ g/cm^3, the neutrino mean free path is about 1.2 km. BBAL found that the trapping density for a supernova process is about $\rho \simeq 0.5\times10^{12}$ g/cm^3. Once the neutrinos are trapped the collapsing stellar core is denied a most efficient cooling mechanism. Consequently, the equation of state of the core material follows an adiabatic trajectory instead of an isothermal trajectory. The average entropy per nucleon along the adiabatic trajectory is determined by that occurring at trapping density. BBAL

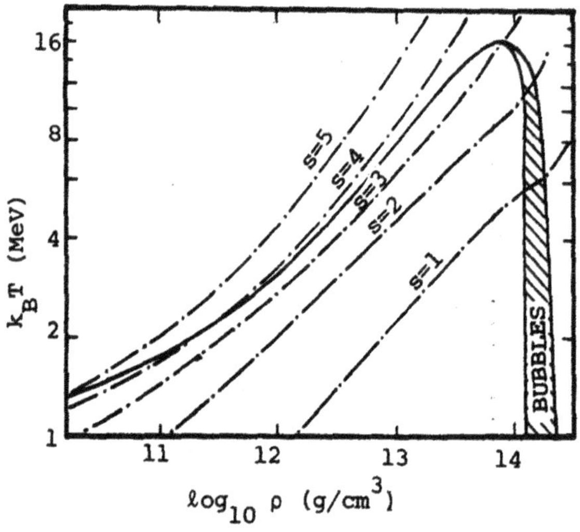

Figure 10-5. Adiabats, labelled by entropy per nucleon, s, in units of k_B for hot dense matter with Y_p=0.25.

estimated the average entropy per nucleon at trapping density to be $(s/k_B) \simeq 1.1$ (Brown, Bethe and Baym 1982). In Figure 10-5 adiabats on the ρ-T plot are shown. This observation by BBAL has profound consequence on the study of the supernova process. It simplifies the construction of the equation of state enormously. The equation of state for subnuclear matter at a fixed $Y_p = 0.3$ for an adiabat with an average entropy per nucleon of $(s/k_B) = 1.5$ is listed in Appendix C. The equation of state for transnuclear matter at $Y_p = 0.3$ and $(s/k_B) = 1.1$ is listed in Appendix C. These results applicable to the supernova process are taken from Brown, Bethe and Baym (1982).

Other developments on the equation of state for subnuclear matter at finite temperatures along the same line are given by Barranco and Buchler (1981) and by Friedman and Pandharipande (1981). These studies employed different forms of nuclear interactions from that used here.

References

Barranco, M. and Buchler, J.R. (1981). Phys. Rev. C24, 1191.

Baym, G., Bethe, H.A. and Pethick, C.J. (1971). Nucl. Phys. A175, 225.

Bethe, H.A. (1968). Phys. Rev. 167, 879.

Bethe, H.A., Brown, G.E., Applegate, J. and Lattimer, J.M. (1979). Nucl. Phys. A324, 487.

Bonche, P. and Vautherin, D. (1981). Nucl. Phys. A372, 496.

Brown, G.E., Bethe, H.A. and Baym, G. (1982). Nucl. Phys. A375, 481.

Brueckner, K.A., Buchler, J.R., Jorna, S. and Lombard, R.J. (1968). Phys. Rev. 171, 1188.

Farine, M., Cote, J. and Pearson, J.M. (1980). Nucl. Phys. A338, 86.

Friedman, B. and Pandharipande, V.R. (1981). Nucl. Phys. A361, 502.

Kohler, H.S. (1976). Nucl. Phys. A250, 301.

Lamb, D.G., Lattimer, J.M., Pethick, C.J. and Ravenhall, D.G. (1981). Nucl. Phys. A360, 459.

Lattimer, J.M. (1981). Ann. Rev. Nucl. Part. Sci. 31, 337.

Lattimer, J.M. and Ravenhall, D.G. (1978). Astrophys. J. 223, 314.

Ravenhall, D.G., Bennett, C.D. and Pethick, C.J. (1972). Phys. Rev. Lett. 28,978.

Regime III: $10^{14} < \rho < 10^{16}$ g/cm^3. Nuclear and Transnuclear Matter.

11. Nucleon-nucleon Scattering and Realistic Nuclear Potentials

In this chapter we shall study matter at and above nuclear densities, which we shall refer to as transnuclear densities. It covers a density range approximately between 2×10^{14} to 10^{16} g/cm^3. There is very little experimental results for matter at transnuclear densities, and therefore the structure of matter at these densities are based largely on theoretical deductions. We shall present several theoretical models which are considered applicable to studies of this type. Some minimum requirements on such models are that (i) they should be well-formulated theoretically, (ii) they base their predictions on realistic inputs, and (iii) they predict successfully the properties of matter at nuclear density for which experimental information is available. The application of such models to investigate matter at transnuclear densities even without direct experimental verification would give a sense of reliability. Thus, models based on effective interactions of the type described in Chapter 2 would not be considered adequate for this purpose since they are constructed on a theoretical framework which is based on low density approximations and are only scaled to reproduce nuclear properties. The extension of such models to much higher densities would inevitably lead to misleading results. We are therefore forced to consider theoretical formulations which are appropriate for high density situations. The models described in this chapter represent a step forward in this direction, and regardless of their actual success in their application they will provide better understandings towards a future successful theory.

In the Hartree-Fock approximation of the many-body wave function, nucleons are assumed to be independent of each other. They are described individually by single-nucleon wave functions. They react to an overall potential which is due to the presence of the others but is otherwise completely ignorant of them. A many-body wave function of this type is called uncorrelated. This type of approximation is believed to be adequate for matter at low density when the Fermi momentum of the system is low, in which case the short-ranged repulsion between nucleons is never deeply probed and the actual interaction experienced by each nucleon never depart greatly from the average interaction employed for the calculation. Whatever slight deviations there may be, the situation could be compensated for by a density dependent term in the effective interaction. This then is the physical justification for such models.

In this chapter we shall review attempts to construct a microscopic theory of nuclear matter. By this we mean a theory based on realistic interactions which each nucleon actually experiences in the system. It will include the effects of correlations as demanded by the many-body theory. A successful microscopic theory could of course explain the origin of the effective interaction and provide a better understanding of the structure of matter.

Due to the technical complications of a full-fledged many-body theory, we shall limit our study to the effects involving two-nucleon correlations. The mathematical details of the many-body problem with two-nucleon correlations resemble in many respects those of the nucleon-nucleon scattering. In preparation for the many-body problem we shall first study the scattering of two nucleons in the Schrodinger representation. The partial wave analysis of the scattering wave function will be presented leading to the construction of realistic two-nucleon potentials, in particular, the Reid potentials, which will serve as inputs for the many-body calculations.

The Reid potentials are based on a potential form resembling (7.21) but include in addition a spin-orbit term v_{SL} as follows:

$$v = v_C + v_S(\vec{\sigma}_1 \cdot \vec{\sigma}_2) + v_T S_{12} + v_{SL} \vec{S} \cdot \vec{L} \tag{11.1}$$

where $\vec{S} = \vec{s}_1 + \vec{s}_2$ is the total spin of the nucleon pair, and $\vec{L} = \vec{r} \times \vec{p}$ with $\vec{p} = -i\hbar\vec{\nabla}$ (gradient in the two-nucleon relative coordinates). The other terms are as those in (7.21) with v_C giving rise to a central force, v_S a spin-dependent force and v_T a tensor force. The isospin-dependent parts have been dropped since we shall treat proton-proton,

proton-neutron and neutron-neutron scattering separately. A two-nucleon system may be decomposed into states of definite angular momenta, or partial waves. The Reid potentials are constructed specifically for low partial waves, for which separate v_C, v_S, v_T and v_{SL} are prescribed. Hence, the Reid potentials are inherently more complicated than (11.1) in that they incorporate implicitly velocity dependence (in addition to the spin-orbit dependence) for at least those low partial waves they are constructed.

For a two-nucleon system, the spins of the individual nucleons will be combined into singlet and triplet spin states. The angular dependence of the two-nucleon wave function will be decomposed into orbital angular momentum states. Due to the presence of the non-central tensor force v_T in the potential, the orbital angular momentum $\vec{L}$ is no longer a constant of the motion, and so only the total angular momentum $\vec{J} = \vec{L} + \vec{S}$ and its z-component J_z are conserved. The potential (11.1) as written is invariant under the interchange of $\vec{x}_1 \leftrightarrow \vec{x}_2$ or $\vec{\sigma}_1 \leftrightarrow \vec{\sigma}_2$, and therefore both the parity P and the symmetry of the total spin states remain constants of the motion. Since the triplet states are symmetric under the interchange of nucleon spins and the singlet states is antisymmetric, the conservation of this symmetry in an interaction implies the absence of transitions between triplet and singlet states, and thus giving rise to the conservation of the total spin $\vec{S}^2$. Therefore, a two-nucleon system may be characterized by the following constants of motion: $\vec{J}^2$, J_z, $\vec{S}^2$ and P, with eigenvalues denoted by $J(J+1)\hbar^2$, $m\hbar$, $S(S+1)\hbar^2$ and ± 1 (called even or odd), respectively. The eigenvalues of $\vec{L}^2$, denoted by $L(L+1)\hbar^2$, need not be conserved. Even parity states are states of even orbital L and odd parity states are states of odd orbital L. Parity conservation implies no transitions from even to odd L states, or vice versa.

Some of the low partial wave states of a two-nucleon system in spectroscopic notations of ${}^{2S+1}L_J$ (and writing the letters S for L=0, P for L=1, D for L=2, and F for L=3, etc.) are:

1. Singlet spin states (S = 0): J = L

    J = 0, L = 0: ${}^1S_0$

    J = 1, L = 1: ${}^1P_1$

    J = 2, L = 2: ${}^1D_2$

2. Triplet spin states (S = 1): J = L+1, L, L-1

$J = 0,\ L = 1:\ \ {}^3P_0$

$J = 1,\ L = 0,\ 2:\ \ {}^3S_1 + {}^3D_1$

$J = 2,\ L = 1:\ \ {}^3P_1$

$J = 2,\ L = 1,\ 3:\ \ {}^3P_2 + {}^3F_2$

$J = 2,\ L = 2:\ \ {}^3D_2$, etc.

Since L is not conserved, there will be mixing between ${}^3S_1$ and ${}^3D_1$ states, and also between ${}^3P_2$ and ${}^3F_2$ states. In other words starting with a pure ${}^3S_1$ state before scattering, some of it will turn into ${}^3D_1$ state after scattering due mainly to the tensor force which is the source of mixing. These are called coupled states and are listed above on the same line with a plus sign in between. The other states are called uncoupled states.

Let us consider the scattering of two nucleons labelled by subscripts "a" and "b". respectively. Define total and relative momenta by:

$$\vec{K} = \vec{k}_a + \vec{k}_b \qquad \text{and} \qquad \vec{k} = \frac{1}{2}(\vec{k}_a - \vec{k}_b) \ , \tag{11.2}$$

and center-of-mass and relative coordinates by:

$$\vec{X} = \frac{1}{2}(\vec{x}_a + \vec{x}_b) \qquad \text{and} \qquad \vec{r} = \vec{x}_a - \vec{x}_b \ . \tag{11.3}$$

The two-nucleon wave function satisfies the Schrodinger equation:

$$\{-\frac{\hbar^2}{2m}(\nabla_a^2 + \nabla_b^2) + v(|\vec{x}_a - \vec{x}_b|)\}\Psi(\vec{x}_a,\vec{x}_b) = E\ \Psi(\vec{x}_a,\vec{x}_b) \tag{11.4}$$

and is separable into:

$$\Psi(\vec{x}_a,\vec{x}_b) = \frac{1}{\sqrt{\Omega}}\, e^{i\vec{K}\cdot\vec{X}}\, \psi(\vec{r}) \ . \tag{11.5}$$

For elastic scattering processes, the center-of-mass motion described by $\exp(i\vec{K}\cdot\vec{X})$ may be ignored and the final results will be independent of $\vec{K}$. However, $\vec{K}$ will play a role if the interacting nucleons were immersed in a many-body system as we shall consider in the next section. For the time being we shall write $\psi(\vec{r})$ without a subscript $\vec{K}$. The relative wave function $\psi(\vec{r})$ satisfies:

$$\{-\frac{\hbar^2}{m}\nabla^2 + v(r)\}\ \psi(\vec{r}) = \varepsilon\ \psi(\vec{r}) \tag{11.6}$$

where $r = |\vec{r}|$, and

$$\varepsilon = E - \frac{\hbar^2 K^2}{4m} \equiv \frac{\hbar^2 k^2}{m} \tag{11.7}$$

with $K = |\vec{K}|$ and $k = |\vec{k}|$, is the difference between E and the center-of-mass kinetic energy. Let us write (11.6) in the form: (with $\hbar$=1)

$$(\ \nabla^2 + k^2\ )\ \psi(\vec{r}) = m\ v(r)\ \ \psi(\vec{r})\ . \tag{11.8}$$

In a time-independent treatment of scattering, the event is described by a steady plane wave, $\psi \sim \exp(i\vec{k}\cdot\vec{r})$, which is the homogeneous solution of (11.8). Even though the time variable has been suppressed, the terms "incoming" and "outgoing" waves are still used as if ψ is retaining a time factor $\exp(-i\omega t)$. We shall choose the z-axis to be in the direction of $\vec{k}$, the x-axis lying in the plane of $\vec{k}$ and $\vec{S}$, and the y-axis perpendicular to the plane. In terms of the Cartesian coordinates just laid down, the $\vec{r}$ vector is expressed in the spherical coordinate variables (r,θ,ϕ) as follows:

$$\vec{r} = (\ \hat{e}_x\ r\ \sin\theta\ \cos\phi,\ \hat{e}_y\ r\ \sin\theta\ \sin\phi,\ \hat{e}_z\ r\ \cos\theta\ ). \tag{11.9}$$

In addition to the homogenous solution we seek a particular solution of ψ from (11.8), which asymptotically represents a spherical outgoing wave as follows:

$$\psi(\vec{r}) \xrightarrow[r\to\infty]{} e^{i\vec{k}\cdot\vec{r}} + f(\theta,\phi)\ \frac{e^{ikr}}{r}\ , \tag{11.10}$$

where $f(\theta,\phi)$ being the amplitude of the outgoing (or scattered) wave is called the scattering amplitude. The intensity of the scattered wave into the solid angle $d\Omega = \sin\theta d\theta d\phi$ is proportional to $|f(\theta,\phi)|^2 d\Omega$. In particular, the differential cross-section of scattering into the solid angle is given by:

$$\frac{d\sigma}{d\Omega} = |f(\theta,\phi)|^2\ , \tag{11.11}$$

and the total cross-section:

$$\sigma = \int |f(\theta,\phi)|^2\ d\Omega\ . \tag{11.12}$$

The general solution to (11.8) may be obtained from the Green's function method. Defining the Green's function $g(\vec{r}-\vec{r}')$ by:

$$(\nabla^2 + k^2) \; g(\vec{r}-\vec{r}') \;=\; \delta(\vec{r}-\vec{r}') \;, \tag{11.13}$$

then,

$$\psi(\vec{r}) \;=\; e^{i\vec{k}\cdot\vec{r}} + \int d^3r' \; g(\vec{r}-\vec{r}') \; m \; v(r') \; \psi(\vec{r}') \;. \tag{11.14}$$

Through a Fourier decomposition of $g(\vec{r}-\vec{r}')$ and $\delta(\vec{r}-\vec{r}')$:

$$g(\vec{r}-\vec{r}') \;=\; (2\pi)^{-3} \int d^3k' \; e^{i\vec{k}'\cdot(\vec{r}-\vec{r}')} \; \tilde{g}(\vec{k}') \;, \tag{11.15}$$

and,

$$\delta(\vec{r}-\vec{r}') \;=\; (2\pi)^{-3} \int d^3k' \; e^{i\vec{k}'\cdot(\vec{r}-\vec{r}')} \;, \tag{11.16}$$

it can easily be deduced that:

$$\tilde{g}(\vec{r}-\vec{r}') \;=\; (-k'^2+k^2)^{-1} \;. \tag{11.17}$$

Thus,

$$\psi(\vec{r}) \;=\; e^{i\vec{k}\cdot\vec{r}} - (2\pi)^{-3} \int d^3r' d^3k' \; \frac{e^{i\vec{k}'\cdot(\vec{r}-\vec{r}')}}{k'^2-k^2-i\varepsilon} \; mv(r')\psi(\vec{r}') \;, \tag{11.18}$$

where an imaginary infinitesimal part, $-i\varepsilon$, is added to the denominator in order to obtain an outgoing wave solution in the asymptotic region as specified by (11.10). Such an asymptotic result can be demonstrated by integrating (11.18) over $\vec{k}'$ as follows. With $d^3k' = k'^2 dk' d\cos\theta' d\phi'$, the relevant integration is:

$$\int d\cos\theta' d\phi' \; e^{ik'|\vec{r}-\vec{r}'|\cos\theta'} \;=\; (2\pi) \; \frac{e^{ik'|\vec{r}-\vec{r}'|} - e^{-ik'|\vec{r}-\vec{r}'|}}{ik'|\vec{r}-\vec{r}'|} \;,$$

Next, the dk' integration may be carried out by first noting that the integration is even in k' and the limits of integration, which are 0 and $+\infty$, may be changed to run from $-\infty$ to $+\infty$. The integration is then converted into a contour integration with the result given by the residual theorem as follows:

$$\psi(\vec{r}) \;=\; e^{i\vec{k}\cdot\vec{r}} - \frac{1}{4\pi} \int d^3r' \; \frac{e^{ik|\vec{r}-\vec{r}'|}}{|\vec{r}-\vec{r}'|} \; m \; v(r')\psi(\vec{r}') \;. \tag{11.19}$$

For $r \gg r'$, since the range of r' is limited by $v(r')$, the phase of the exponential may be approximated by:

$$k|\vec{r}-\vec{r}'| = kr \left(1 + \left(\frac{r'}{r}\right)^2 - \frac{2\vec{r}\cdot\vec{r}'}{r^2}\right)^{1/2} \;\simeq\; kr - \vec{k}'\cdot\vec{r}' \tag{11.20}$$

where $\vec{k}' = k\vec{r}/r$, and the denominator by $|\vec{r}-\vec{r}'| \approx r$, then (11.19) reduces to:

$$\psi(\vec{r}) \xrightarrow[r\to\infty]{} e^{i\vec{k}\cdot\vec{r}} - \left(\frac{e^{ik\cdot r}}{r}\right) \frac{1}{4\pi} \int d^3r' \, e^{-i\vec{k}'\cdot\vec{r}'} mv(r')\psi(\vec{r}'). \qquad (11.21)$$

Comparing (11.10) with (11.21), the scattering amplitude is identified to be:

$$f(\theta,\phi) = -(4\pi)^{-1}\int d^3r' \, e^{-i\vec{k}'\cdot\vec{r}'} mv(r')\psi(\vec{r}') \equiv f(\hat{k},\hat{k}') \;, \qquad (11.22)$$

where the angular variables (θ,ϕ) correspond to the direction of $\hat{k}'$ with respect to $\hat{k}$, and $f(\theta,\phi)$ is written specificallyas $f(\hat{k},\hat{k}')$, which will be useful later. An integral equation for the scattering amplitude may be obtained if we integrate over (11.18) on both sides, i.e.,

$$-(4\pi)^{-1}\int d^3r \, e^{-i\vec{k}'\cdot\vec{r}} mv(r\,)\psi(\vec{r}) = -(4\pi)^{-1}\int d^3r \, e^{i(\vec{k}-\vec{k}')\cdot\vec{r}} \, mv(r) \; +$$

$$-(2\pi)^{-3}\int d^3k'' \frac{1}{k''^2-k^2-i\varepsilon} \, d^3r \, e^{i(\vec{k}''-\vec{k}')\cdot\vec{r}} mv(r)\{-(4\pi)^{-1}\int d^3r' e^{-i\vec{k}''\cdot\vec{r}'} mv(r')\psi(r')\}$$

or,

$$f(\hat{k},\hat{k}') = -(4\pi)^{-1} \tilde{U}(\vec{k}-\vec{k}') - (2\pi)^{-3}\int d^3k'' \frac{\tilde{U}(\vec{k}''-\vec{k}')}{k''^2-k^2-i\varepsilon} f(\hat{k}',\hat{k}''), \quad (11.23)$$

where $\tilde{U}(\vec{k})$ is the Fourier transform of $mv(r)$ and $-\tilde{U}(\vec{k}-\vec{k}')/4\pi$ is the Born approximation to $f(\hat{k},\hat{k}')$. (11.18) or (11.23) is sometimes called the Lippmann-Schwinger equation.

In the case of the nucleon scattering, let us first consider the two-nucleon system in the singlet spin state. This situation is particularly simple, and with S=0 the potential (11.1) acting on the spin wave function simplifies to:

$$\begin{aligned} v\,\chi_0 &= (\, v_C + v_S\vec{\sigma}_1\cdot\vec{\sigma}_2 + v_T S_{12} + v_{SL}\vec{S}\cdot\vec{L}\,)\,\chi_0 \\ &= (\, v_C - 3v_S)\,\chi_0 \;, \end{aligned} \qquad (11.24)$$

which depends only on the magnitude of the relative coordinate $\vec{r}$. The scattering process possesses also azimuthal symmetry and is independent of the variable ϕ. We note that for proton-proton scattering a Coulomb term should be added to the potential; we shall assume that this is understood.

The partial wave decomposition of ψ is given by:

$$\psi(\vec{r}) = \sum_{L=0}^{\infty} (2L+1)\, i^L \, \frac{u_L(kr)}{kr} \, P_L(\hat{k}\cdot\hat{r})\, \chi_0 \;, \qquad (11.25)$$

where P_L are the Legendre polynomials (see Appendix B.2), and u_L satisfies:

$$\frac{d^2u_L}{dr^2} + \left\{k^2 - \frac{L(L+1)}{r^2} - mv(r)\right\} u_L(kr) = 0 , \qquad (11.26)$$

according to (11.6), with v(r) given by (11.24). $u_L(r)$ are further required to satisfy boundary conditions, which are that $u_L(0) = 0$, and that asymptotically they approach the Lth partial waves of a plane wave. It is more satisfactory to work with an equivalent expression for u_L in the form of an integral equation and thereby incorporating the boundary conditions in a single statement. It can be obtained from a partial wave decomposition of (11.18). We shall need the expression for the partial wave expansion of a plane wave:

$$\begin{aligned} e^{i\vec{k}\cdot\vec{r}} &= \sum_{L=0}^{\infty} (2L+1)\, i^L\, j_L(kr)\, P_L(\hat{k}\cdot\hat{r}) \\ &= 4\pi \sum_{L=0}^{\infty} \sum_{M=-L}^{L} i^L\, j_L(kr)\, Y^*_{LM}(\hat{k})\, Y_{LM}(\hat{r}) , \end{aligned} \qquad (11.27)$$

where $j_L(kr)$ are the spherical Bessel functions (see Appendix B.3). (11.18) is expanded into:

$$\begin{aligned} &\sum_L (2L+1)\, i^L \left(\frac{u_L(kr)}{kr} - j_L(kr)\right) P_L(\hat{k}\cdot\hat{r}) = \\ &= -\frac{(4\pi)^3}{(2\pi)^3}\int d^3k'\int d^3r'\Big\{\sum_{\ell,m} i^\ell j_\ell(k'r)Y^*_{\ell m}(\hat{k}')Y_{\ell m}(\hat{r})\Big\}\Big\{\sum_{\ell' m'}(-i)^{\ell'} j_{\ell'}(k'r')Y_{\ell' m'}(\hat{k}')Y^*_{\ell' m'}(\hat{r}')\Big\}\times \\ &\quad \times\Big\{\sum_{\ell''}(i)^{\ell''}\frac{u_{\ell''}(kr')}{kr'}Y_{\ell''0}(\hat{r}')\big((2\ell''+1)/4\pi\big)^{1/2}\Big\}\frac{mv(r')}{k'^2-k^2-i\varepsilon} \\ &= -(2/\pi)\int_0^\infty k'^2dk'\int_0^\infty r'^2dr'\sum_L (2L+1)i^L\,\frac{j_L(k'r)j_L(k'r')}{k'^2-k^2-i\varepsilon}mv(r')\frac{u_L(kr')}{kr'}P_L(\hat{k}\cdot\hat{r}) . \end{aligned}$$

(11.28)

The partial wave amplitudes are decoupled into the following set of integral equations:

$$\frac{u_L(kr)}{kr} = j_L(kr) - \frac{2}{\pi}\int_0^\infty k'^2dk'\int_0^\infty r'^2dr'\,\frac{j_L(k'r)j_L(k'r')}{k'^2-k^2-i\varepsilon}\,mv(r')\frac{u_L(kr')}{kr'} ,$$

(11.29)

which may be written in a more compact form as:

$$u_L(kr) = \mathcal{J}_L(kr) - \frac{2}{\pi}\int_0^\infty dk'\int_0^\infty dr'\frac{\mathcal{J}_L(k'r)\mathcal{J}_L(k'r')}{k'^2-k^2-i\varepsilon}mv(r')\, u_L(kr'), \qquad (11.30)$$

where $J_L(kr) = kr\, j_L(kr)$, and they satisfy:

$$\left\{ \frac{d^2}{dr^2} - \frac{L(L+1)}{r^2} + k^2 \right\} J_L(kr) = 0 . \qquad (11.31)$$

(11.29) can further be reduced by carrying out the k' integration. Since the integrand is even in k', the limits of integration may be changed so as to run from $-\infty$ to $+\infty$ and the integral is converted into a contour integral. $j_L(z)$ is an entire function of z and the integrand has only simple poles at $k' = \pm(k + i\varepsilon)$. Consider first the case $r>r'$. Let us replace $j_L(k'r)$ by $\frac{1}{2}\left(h_L(k'r) + h_L^*(k'r)\right)$ where h_L are the Hankel functions (see Appendix B.3). Then $j_L(k'r')h_L(k'r)$ would decrease exponentially in the upper complex k'-plane as $|k'| \to \infty$. (The asymptotic forms of h_L are given by B.3.10). The integral may be evaluated by the residual method with the contour closed in the upper half-plane. Similarly $j_L(k'r')h_L^*(k'r)$ term is evaluated by closing the contour in the lower half-plane. (11.29) is evaluated to be:

$$\frac{u_L(kr)}{kr} = j_L(kr) - i\int_0^\infty r'dr'\, j_L(kr_<)h_L(kr_>)mv(r')u_L(kr'), \qquad (11.32)$$

where $r_<$ ($r_>$) is the smaller (larger) of r and r'.

Before we proceed any further, let us summarize the different forms of the Green's function $g(\vec{r}-\vec{r}\,')$ we have obtained so far:

$$g(\vec{r}-\vec{r}\,') = -(2\pi)^{-3}\int d^3k' \frac{e^{i\vec{k}'\cdot(\vec{r}-\vec{r}\,')}}{k'^2-k^2-i\varepsilon} = -\frac{1}{4\pi}\frac{e^{ik|\vec{r}-\vec{r}\,'|}}{|\vec{r}-\vec{r}\,'|}$$

$$= \sum_{L=0}^{\infty}\sum_{M=-L}^{L} g_L(r,r')\, Y_{LM}(\hat{r})\, Y^*_{LM}(\hat{r}') , \qquad (11.33)$$

where,

$$g_L(r,r') = -\frac{1}{\pi}\int_{-\infty}^{\infty} k'^2dk' \frac{j_L(k'r)j_L(k'r')}{k'^2-k^2-i\varepsilon} = -ik\, j_L(kr_<)h_L(kr_>) .$$

The asymptotic form for $g_L(r,r')$ is:

$$g_L(r,r') \xrightarrow[r\to\infty]{} -\frac{e^{ikr}}{r}(-i)^L j_L(kr') .$$

Thus, (11.29) or (11.32) provides an evaluation of u_L once the potential v is given. On the other hand, if we are interested in the construction of v from the scattering data,we must relate v to the scattering amplitude $f(\theta)$, which is directly related to the scattering cross-section. With $f(\theta)$ expanded into partial waves in a similar manner;

$$f(\theta) = \sum_L (2L+1)\, f_L(k)\, P_L(\hat{k}\cdot\hat{r}) \ , \tag{11.34}$$

(11.21) is written as;

$$\psi \to \sum_L (2L+1) i^L \left(j_L(kr) + (-i)^L \frac{e^{ikr}}{r} f_L(k) \right) P_L(\hat{k}\cdot\hat{r}) \ . \tag{11.35}$$

Making use of the asymptotic form of h_L, (11.32) has the asymptotic limit:

$$\frac{u_L(kr)}{kr} \to j_L(kr) - i\,\frac{e^{ikr}}{r}\, e^{-\frac{i}{2}(L+1)\pi} \int_0^\infty r'dr' j_L(kr') mv(r') u_L(kr') . \tag{11.36}$$

Comparing (11.35) with (11.36), it is easy to identify that:

$$f_L(k) = -\int_0^\infty r'dr'\ j_L(kr')\ mv(r')\ u_L(kr') \ . \tag{11.37}$$

The construction of potential from scattering data proceeds usually in two steps: the first step is to determine from the data sets of phase shifts for different partial waves, and the second step is to relate the potential v to the phase shifts.

The construction of potential from scattering data is a tedious process. We shall not carry the analysis out any further. The interested readers are referred to the standard text by Wu and Ohmura (1962). Instead, let us turn to the case involving triplet spin states.

For triplet spin states, the orbital states are combined with the spin states into expressions called the vector spherical harmonics:

$$\mathcal{Y}^M_{J1L}(\theta,\phi) = \sum_m (JM|L\ M-m, 1m)\ Y_{L,M-m}(\theta,\phi)\ \chi_1^m \ , \tag{11.38}$$

where $(JM|Lm_L, Sm_S)$ are the Clebsch-Gordan coefficients with S=1. Explicitly,

$$\mathcal{Y}^M_{J1J-1} = \left(2J(2J-1)\right)^{-1/2} \{ \left((J+M-1)(J+M) \right)^{1/2} Y_{J-1,M-1}\, \chi_1^1 + $$
$$+ \left(2(J-M)(J+M)\right)^{1/2} Y_{J-1,M}\, \chi_1^0 + \left((J-M-1)(J-M)\right)^{1/2} Y_{J-1,M+1}\, \chi_1^{-1} \},$$

$$\mathcal{Y}^M_{J1J} = \left(2J(J+1)\right)^{-1/2} \{ -\left((J+M)(J-M+1)\right)^{1/2} Y_{J,M-1}\, \chi_1^1 + $$
$$+ \sqrt{2}\, M\, Y_{J,M}\, \chi_1^0 + \left((J-M)(J+M+1)\right)^{1/2} Y_{J,M+1}\, \chi_1^{-1} \} \ ,$$

$$\mathcal{Y}^M_{J1J+1} = \left(2(J+1)(2J+3)\right)^{-1/2} \{ \left((J-M+1)(J-M+2)\right)^{1/2} Y_{J+1,M-1}\, \chi^1_1 +$$

$$-\left(2(J-M+1)(J+M+1)\right)^{1/2} Y_{J+1,M}\, \chi^0_1 + \left((J+M+2)(J+M+1)\right)^{1/2} Y_{J+1,M+1} \chi^{-1}_1 \}$$

(11.39)

$\mathcal{Y}^M_{JSL}$ are orthogonal functions in the sense that:

$$\int d\Omega\ \mathcal{Y}^{M^*}_{JSL}(\theta,\phi)\ \mathcal{Y}^{M'}_{J'S'L'}(\theta,\phi) = \delta_{JJ'}\delta_{SS'}\delta_{LL'}\delta_{MM'} \ . \tag{11.40}$$

In terms of the vector spherical harmonics, the partial wave decomposition of ψ is given by:

$$\psi = \sum_{J=0} \sum_{L=J,J\pm 1} (2J+1) i^J \frac{w_{JL}(kr)}{kr} \mathcal{Y}^M_{J1L}(\theta,\phi) \ . \tag{11.41}$$

Acting on the triplet spin state the different terms of the potential (11.1) yield the following results. Some are relatively simple:

$$\{ v_C + v_S(\vec{\sigma}_1\cdot\vec{\sigma}_2)\}\ \chi^m_1 = (v_C + v_S)\ \chi^m_1 \ ,$$
$$\vec{L}\cdot\vec{S}\ \mathcal{Y}^M_{J1L} = \frac{1}{2}\{J(J+1) - L(L+1) - 2\}\mathcal{Y}^M_{J1L} \ . \tag{11.42}$$

For the tensor operator part more analysis is needed. Let us first rewrite S_{12} in terms of $\vec{S}= \frac{1}{2}(\vec{\sigma}_1+\vec{\sigma}_2)$:

$$S_{12} = 3(\vec{\sigma}_1\cdot\hat{r})(\vec{\sigma}_2\cdot\hat{r}) - (\vec{\sigma}_1\cdot\vec{\sigma}_2) = 6(\vec{S}\cdot\hat{r})^2 - 2S^2 = 6(\vec{S}\cdot\hat{r})^2 - 4, \tag{11.43}$$

where the last line is written with explicitly the triplet states in mind, for which $\vec{S}^2\ \chi^m_1 = 2\ \chi^m_1$. Define,

$$S_\pm = (S_x \pm iS_y) \ , \tag{11.44}$$

where S_+ (S_-) is the raising (lowering) operator:

$$S_\pm\ \chi^m_1 = \sqrt{(1 \mp m)(2 \pm m)}\ \chi^{m\pm 1}_1 \ . \tag{11.45}$$

Let us rewrite:

$$(\vec{S}\cdot r) = \frac{1}{r}\{z\ S_z + \frac{1}{2}(x-iy)S_+ + \frac{1}{2}(x+iy)S_-\}$$

$$= \cos\theta\ S_z + \frac{1}{2}\sin\theta\ e^{-i\phi}S_+ + \frac{1}{2}\sin\theta\ e^{i\phi}S_- \ , \tag{11.46}$$

according to (11.9). Thus,

$$(\vec{S}\cdot\hat{r})\ \chi^1_1 = \cos\theta\ \chi^1_1 + \frac{1}{\sqrt{2}}\sin\theta\ e^{i\phi}\chi^0_1 \ ,$$

$$(\vec{S}\cdot\hat{r})\,\chi_1^0 = \frac{1}{\sqrt{2}}\sin\theta\, e^{-i\phi}\,\chi_1^1 + \frac{1}{\sqrt{2}}\sin\theta\, e^{i\phi}\,\chi_1^{-1}\,,$$

$$(\vec{S}\cdot\hat{r})\,\chi_1^{-1} = \frac{1}{\sqrt{2}}\sin\theta\, e^{-i\phi}\,\chi_1^0 - \cos\theta\,\chi_1^{-1}\,, \tag{11.47}$$

or,

$$S_{12}\,\chi_1^1 = (3\cos^2\theta-1)\chi_1^1 + 3\sqrt{2}\,\sin\theta\cos\theta e^{i\phi}\chi_1^0 + 3\,\sin^2\theta e^{2i\phi}\chi_1^{-1}\,,$$

$$S_{12}\,\chi_1^0 = 3\sqrt{2}\,\sin\theta\cos\theta e^{-i\phi}\chi_1^1 + (6\sin^2\theta-4)\chi_1^0 - 3\sqrt{2}\,\sin\theta\cos\theta e^{i\phi}\chi_1^{-1}\,,$$

$$S_{12}\,\chi_1^{-1} = 3\,\sin^2\theta e^{-2i\phi}\chi_1^1 + 3\sqrt{2}\,\sin\theta\cos\theta e^{-i\phi}\chi_1^0 + (3\cos^2\theta-1)\chi_1^{-1}. \tag{11.48}$$

Consequently, the acticn of S_{12} on the vector spherical harmonics behaves as follows. For the J=L states, parity conservation forbids their transitions to J=L±1 states, and they belong to the uncoupled states:

$$S_{12}\,\mathcal{Y}_{J1J}^M = a_J\,\mathcal{Y}_{J1J}^M\,, \tag{11.49}$$

where a_J is independent of M. a_J can most easily be evaluated in the special case with $\theta=0$, so that the spherical harmonics $Y_{Jm}(0)=\sqrt{(2J+1)/4\pi}\,\delta_{m0}$ and for M=1:

$$\mathcal{Y}_{J1J}^M = -\frac{1}{\sqrt{2}}\,Y_{J0}(0)\,\chi_1^1\,.$$

Together with the fact that $S_{12}\chi_1^1 = 2\,\chi_1^1$, we find that:

$$a_J = 2\,. \tag{11.50}$$

On the other hand, S_{12} couples the J=L±1 states. Let us write:

$$S_{12}\,\mathcal{Y}_{J1J-1}^M = a_{--}\,\mathcal{Y}_{J1J-1}^M + a_{-+}\,\mathcal{Y}_{J1J+1}^M\,,$$

$$S_{12}\,\mathcal{Y}_{J1J+1}^M = a_{+-}\,\mathcal{Y}_{J1J-1}^M + a_{++}\,\mathcal{Y}_{J1J+1}^M\,, \tag{11.51}$$

where the a's are also independent of M. Again, specialize the results to $\theta=0$, and for M=0 and M=1, it can then be found that:

$$a_{--} = -\frac{2(J-1)}{2J+1}\,,$$

$$a_{+-} = a_{-+} = \frac{6\sqrt{J(J+1)}}{2J+1}\,,$$

$$a_{++} = -\frac{2(J+2)}{2J+1}\,. \tag{11.52}$$

The Schrodinger equation for each partial wave becomes:

J=L: $$\left\{\frac{d^2}{dr^2} + k^2 - \frac{J(J+1)}{r^2} - v_J^J\right\} w_{JJ} = 0\,,$$

$J = L\pm 1$:

$$\left[\frac{d^2}{dr^2} + k^2 - \frac{J(J-1)}{r^2} - v^J_{--}\right] w_{J,J-1} = v^J_{-+} w_{J,J+1} ,$$

$$\left[\frac{d^2}{dr^2} + k^2 - \frac{(J+1)(J+2)}{r^2} - v^J_{++}\right] w_{J,J+1} = v^J_{+-} w_{J,J-1} , \tag{11.53}$$

where,

$$v^J_J = m\left(v_C + v_S - v_{SL} + 2v_T\right) ,$$

$$v^J_{--} = m\left(v_C + v_S + (J-1)v_{SL} - \frac{2(J-1)}{2J+1} v_T\right) ,$$

$$v^J_{+-} = v^J_{-+} = m \frac{6\sqrt{J(J+1)}}{2J+1} v_T ,$$

$$v^J_{++} = m\left(v_C + v_S - (J+2)v_{SL} - \frac{2(J+2)}{2J+1} v_T\right) . \tag{11.54}$$

The connections of these partial waves to phase shifts, and the problem of finding eigenphase shifts will not be discussed. They are explained in detail in Wu and Ohmura (1962). We shall took at a different formulation which is closer related to the correlation problem that we shall encounter in the next section.

A more detailed description of the process may be obtained by decomposing the coupled states in a way which specifies the incoming and outgoing states. Note that an incoming plane wave in a triplet spin state is decomposed into:

$$e^{i\vec{k}\cdot\vec{r}} \chi^m_1 = \sum_{J=0}^{\infty} \sum_{\ell=J,J\pm1} \sqrt{4\pi(2\ell+1)}\, i^\ell (Jm|\ell 0 1m) j_\ell(kr) \mathcal{Y}^m_{J1\ell}(\hat{k}\cdot\hat{r}) , \tag{11.55}$$

where we have made use of (11.27), written $Y^*_{\ell m}(\hat{r}\cdot\hat{r}) = \sqrt{(2\ell+1)/4\pi}\,\delta_{m0}$ and combined $Y_{\ell 0}(\hat{k}\cdot\hat{r})$ with χ^m_1 to form $\mathcal{Y}^m_{J1\ell}(\hat{k}\cdot\hat{r})$ according to (11.38). In a similar way, we may decompose ψ into:

$$\psi = \sum_{J=0}^{\infty} \sum_{\ell=J,J\pm1} \sqrt{4\pi(2\ell+1)}\, i^\ell (Jm|\ell 0 1m) \sum_{\ell'=J,J\pm1} \frac{w^\ell_{J,\ell'}}{kr} \mathcal{Y}^m_{J1\ell}(\hat{k}\cdot\hat{r}) , \tag{11.56}$$

where ℓ correspond to the orbital angular momenta of the incoming states (also referred to as the entrance channels) and ℓ' the outgoing states (exit channels). Superscripts ℓ are added to the radial wave functions $w_{J\ell'}$ to identify the sources from which they originate. This decomposition is clearly permissible and it retains the maximum resemblance to the decomposition of the incoming plane wave.

To obtain the integral equation formulation of the scattering event, we insert (11.5) and (11.56) into (11.14) to find:

$$\sum_{\ell,\ell'=J,J\pm1} \sqrt{(2\ell+1)}\; i^{\ell}(Jm|\ell 01m)\{\, w^{\ell}_{J\ell'}\, \mathcal{Y}^{m}_{J1\ell'}(\hat{k},\hat{r}) - kr\, j_{\ell}\, \mathcal{Y}^{m}_{J1\ell}(\hat{k},\hat{r})\} =$$

$$= \int d^3r'\,(r/r')\{\sum_{\ell'',m''} g_{\ell''}(r,r')Y_{\ell''m''}(\hat{r})Y^{*}_{\ell''m''}(\hat{r}')\}mv(r')\times$$

$$\times \sum_{\ell,\ell'=J,J\pm1} \sqrt{(2\ell+1)}\; i^{\ell}(Jm|\ell 01m)w^{\ell}_{J\ell'}\, \mathcal{Y}^{m}_{J1\ell'}(\hat{k},\hat{r})$$

$$= \sum_{\ell,\ell',\ell''=J,J\pm1} \sqrt{(2\ell+1)}\; i^{\ell}(Jm|\ell 01m)\; r\int_0^{\infty} r'dr'\; g_{\ell''}(r,r')w^{\ell}_{J\ell'}(v)_{\ell'\ell''}\, \mathcal{Y}^{m}_{J1\ell''}(\hat{k},\hat{r}) \tag{11.57}$$

where $(v)_{\ell'\ell''}$ represent the matrix elements $\langle \mathcal{Y}^{m}_{J1\ell'}|mv(r')|\mathcal{Y}^{m}_{J1\ell''}\rangle$ as given by (11.54), and the last line is obtained after an integration over the angular variables of r' has been performed. A particular (J,L) component can now be projected out of (11.57) giving:

$$\sum_{\ell}\sqrt{(2\ell+1)}\; i^{\ell}(Jm|\ell 01m)\; w^{\ell}_{JL} = \sqrt{(2J+1)}\; i^{L}(Jm|L01m)\; kr\; j_L \; +$$

$$+ \sum_{\ell,\ell'} \sqrt{(2\ell+1)}\; i^{\ell}(Jm|\ell 01m)\; r\int_0^{\infty} r'dr'\; g_L(r,r')\; w^{\ell}_{J\ell'}\; (v)_{\ell' L}\,. \tag{11.58}$$

Finally, the component with definite L' in the entrance channel can be selected by multiplying (11.58) by (Jm|L'01m) and sum over m making use of the relation:

$$\sum_{m=\pm1,0} (Jm|L'01m)(Jm|\ell 01m) = \delta_{L'\ell}\,\frac{2J+1}{2L'+1}\,. \tag{11.59}$$

Thus,

$$w^{L'}_{JL}(kr) = \delta_{LL'}\; kr\; j_L(kr) + r\int_0^{\infty} r'dr'\; g_L(r,r')\{\sum_{\ell} w^{L'}_{J\ell}\; (v)_{\ell L}\}, \tag{11.60}$$

or, explicitly,

J=L: $$w^{J}_{J,J} = kr\; j_J + r\int_0^{\infty} r'dr'\; g_J(r,r')v_J(r')w^{J}_{JJ}(kr')\,,$$

J=L±1: $$w^{J-1}_{J,J-1} = kr\; j_{J-1} + r\int_0^{\infty} r'dr'\; g_{J-1}(r,r')\{v_{--}w^{J-1}_{J,J-1} + v_{+-}w^{J-1}_{j,J+1}\},$$

$$w^{J+1}_{J,J+1} = kr\; j_{J+1} + r\int_0^{\infty} r'dr'\; g_{J+1}(r,r')\{v_{++}w^{J+1}_{J,J+1} + v_{-+}w^{J+1}_{J,J-1}\},$$

$$w^{J+1}_{J,J-1} = r\int_0^{\infty} r'dr'\; g_{J-1}(r,r')\{\, v_{--}w^{J+1}_{J,J-1} + v_{+-}w^{J+1}_{J,J+1}\}\,,$$

$$w^{J-1}_{J,J+1} = r\int_0^\infty r'dr'\, g_{J+1}(r,r')\{ v_{++}w^{J-1}_{J,J+1} + v_{-+}w^{J-1}_{J,J-1}\}. \quad (11.61)$$

These are the basic equations for the determination of the partial wave radial wave functions once the interaction potentials are given. These relations are also basic to Brueckner's theory of the many-body problem as we shall describe in the next section.

The problem of extracting nuclear potentials from scattering data is a difficult task. We shall not go into the detailed steps of such a problem. The Reid potentials (Reid 1968) which we shall describe are sets of potentials which when applied to (11.61) will reproduce the experimental phase shifts approximately. There are two distinct types of the Reid potentials known as the Reid hard core and the Reid soft core. The hard core potentials are characterized by an infinite repulsive core of 0.21 fm^{-1} radius, except for the $^1S_0$ state, which is sensitive to the hard core radius and for which the hard core radius is adjusted to be 0.2073 fm^{-1}. The soft core potentials consist of Yukawa potential functions of different ranges and are not prescribed by hard cores.

If nuclear potentials are in general static with the possible addition of a spin-orbit term, then a single set of v(r) would suffice for all partial waves. Reid, however, did not find this to be convenient to accomplish. He and many investigators before him found that the $^1S_0$ and $^1D_2$ phase shifts could not be fitted by a single local potential v(r) (as the spin-orbit term does not enter for the singlet spin state). It seems that velocity dependence beyond (11.1) is needed. Many nuclear potentials with different forms of velocity dependence have been constructed (see Review in Bethe 1971). Reid adopted a solution to this situation which is probably the simplest of all. He incorporates velocity dependence in the potentials by constructing separate potentials for individual partial wave states. This is done for all the states listed explicitly in the beginning of this section. For higher partial wave states, which interact at longer range, the potentials deduced from meson theory due to one-pion-exchange is to be used. Charge independence in nuclear forces is assumed since there is very little neutron-neutron scattering results. We list here the Reid soft core potentials which are in units of MeV. The following parameters are used: $x = \mu r$ with $\mu = 0.7\ \text{fm}^{-1}$ and $H = 10.463$ MeV.

(a) Isospin T = 1:

$$v(^1S_0) = -H(e^{-x}/x) - 1650.6(e^{-4x}/x) + 6482.2(e^{-7x}/x)$$

$$v(^1D_2) = -H(e^{-x}/x) - 12.322(e^{-2x}/x) - 1112.6(e^{-4x}/x) + 6484.2(e^{-7x}/x)$$

$$v(^3P_0) = -H(1 + 4/x + 4/x^2)e^{-x} - (16/x^2 + 4/x^3)e^{-4x} + \\ +27.133(e^{-2x}/x) - 790.74(e^{-4x}/x) + 20662(e^{-7x}/x)$$

$$v(^3P_1) = H(1 + 2/x + 2/x^2)e^{-x} - (8/x^2 - 2/x^3)e^{-4x} + \\ -135.25(e^{-2x}/x) + 472.81(e^{-3x}/x)$$

$$v(^3P_2+^3F_2) = v_C + v_T\ S_{12} + v_{LS}\ \vec{L}\cdot\vec{S}$$

where,

$$v_C = (H/3)(e^{-x}/x) - 933.48(e^{-4x}/x) + 4152.1(e^{-6x}/x)$$

$$v_T = H(1/3 + 1/x + 1/x^2)e^{-x} - (4/x^2 + 1/x^3)e^{-4x} - 34.925(e^{-3x}/x)$$

$$v_{LS} = -2074.1(e^{-6x}/x)$$

(b) Isospin T = 0

$$v(^1P_1) = 3H(e^{-x}/x) - 634.39(e^{-2x}/x) + 2163.4(e^{-3x}/x)$$

$$v(^3D_2) = -3H(1 + 2/x + 2/x^2)e^{-x} - (8/x^2 + 2/x^3)e^{-4x} + \\ -220.12(e^{-2x}/x) + 871(e^{-3x}/x)$$

$$v(^3S_1+^3D_1) = v_C + v_T\ S_{12} + v_{LS}\ \vec{L}\cdot\vec{S}$$

where,

$$v_C = -H(e^{-x}/x) + 105.468(e^{-2x}/x) - 3187.8(e^{-4x}/x) + 9924.3(e^{-6x}/x)$$

$$v_T = -H(1 + 3/x + 3/x^2)e^{-x} - (12/x^2 + 3/x^3)e^{-4x} + \\ + 351.77(e^{-4x}/x) - 1673.5(e^{-6x}/x)$$

$$v_{LS} = 708.91(e^{-4x}/x) - 2713.1(e^{-6x}/x)$$

For the uncoupled states, there is no need to break up the potential function into parts. For the coupled states, however, the potential function must carry the tensor operator (for mixing) and the spin-orbit operator (for L-dependence) with it. The spin-dependent part may be absorbed into v_C. The Reid soft core potentials given here can reproduce experimental phase shifts up to scattering energies of 350 MeV (laboratory). There are other nuclear potentials expressed usually in somewhat different forms. References to many of these may be found in Bethe's review article (1971) on nuclear matter. The more recent potentials are those of the Paris group (Lacombe *et al*. 1980) and the Bonn group (Holinde and Machleidt 1975).

These potentials provide the needed inputs to a computation of nuclear matter properties. This we shall describe in the next section.

References

Bethe, H.A. (1971). Ann. Rev. Nucl. Sci. **21**, 93.

Holinde, K. and Machleidt, R. (1975). Nucl. Phys. **A247**, 495.

Lacombe, M., Loiseau, B., Richard, J.M., Vinh Mau, R., Cote, J., Pires, P. and de Towrreil, R. (1980). Phys. Rev. **C21**, 861.

Reid, R.V. (1968). Ann. Phys. **50**, 411.

Wu, Ta-You and Ohmura, T. (1962). *Quantum Theory of Scattering*, Prentice-Hall, N.J.

Bibliography

Goldberger, M.L. and Watson, K.M, (1964). *Collision Theory*, Wiley, New York.

Gottfried, K. (1966). *Quantum Mechanics*, Bejamin, New York.

Rodberg, L.S, and Thaler, R.M. (1967). *Introduction to the Quantum Theory of Scattering*, Academic Press, New York.

12. Independent-pair Model of Nuclear Structure

In this section we shall study an attempt at a microscopic theory of nuclear structure. It is predicated on the supposition that nuclear properties can be explained by realistic nuclear potentials which as described in the last section are determined from nuclear scattering data. Due to the complexity of a full-fledged many-body theory it is understandable that simplifying approximations within such a theory should be taken. An immediate improvement over the uncorrelated Hartree-Fock approach is to take into consideration two-nucleon correlation effects. This is called the independent-pair model. The justification of the model is based on certain assumptions, the validity of which may be examined a posteriori. We shall concentrate on the application of this model to the study of nuclear matter.

The independent-pair model of nuclear matter as formulated by Brueckner (1954), Bethe (1956), and Goldstone (1957) has been extensively discussed in several references (see Day 1967, 1978, and Bethe 1971 for reviews). We shall not go into a general discussion of such a model, since to do so would require a very lengthy description. We shall merely give a heuristic derivation of its basic formulation, called the Brueckner-Bethe-Goldstone (BBG) equation, to high light the physics of the problem involved. Our emphasis is on the mathematical details of the application of the BBG equation to the nuclear matter problem.

Nuclear matter is in principle a medium of infinite extend, but it may be treated as a medium in a finite volume consisting of A nucleons obeying periodic boundary conditions. The coordinates of the A nucleons in the volume will be labelled by coordinates $\vec{x}_1, \vec{x}_2, \ldots .$ $\vec{x}_A$. Recall that in the Hartree-Fock approach the total wave function of the system is represented by a Slater determinant of single-nucleon wave functions. The independent-pair model is an extension of this by considering the interaction between any pair of particles as exactly as possible but neglecting any direct interaction with the others.

Let us consider a model wave function as follows:

$$\Psi_{ab} = A\{ \psi_{ab}(\vec{x}_1,\vec{x}_2)\psi_m(\vec{x}_3) \; . \; . \; . \; . \; . \; \psi_n(\vec{x}_A)\} \quad , \tag{12.1}$$

where A indicates that all terms formed by antisymmetric exchanges of

$\vec{x}_1, \vec{x}_2, \ldots \vec{x}_A$ are to be taken. ψ_m, ψ_n, etc are uncorrelated single-nucleon states as in the Hartree-Fock case, while ψ_{ab} refers to the correlated state of a pair of nucleons in states a and b. Ψ_{ab} singles out a pair of states and is assumed to satisfy the many-body Schrodinger equation:

$$H\,\Psi_{ab} = E_{ab}\,\Psi_{ab}\,. \tag{12.2}$$

Let us study the propertiies of Ψ_{ab}. In the case of nuclear matter Ψ_{ab} is translational invariant, which in turn means that the uncorrelated single-nucleon states are individually translational invariant and are thereby given by plane waves. ψ_m, ψ_n, etc are labelled by quantum numbers consisting of momenta, spins and isospins. The correlated pair ψ_{ab} will also possess a total momentum which is conserved and ψ_{ab} may be expanded into plane wave states as follows:

$$\psi_{ab}(\vec{x}_1,\vec{x}_2) = \psi_a(\vec{x}_1)\psi_b(\vec{x}_2)+\sum_\alpha a_\alpha\psi_\alpha(\vec{x}_1)\psi_b(\vec{x}_2)+\sum_\beta a_\beta\psi_a(\vec{x}_1)\psi_\beta(\vec{x}_2)+\sum_{\alpha,\beta}a_{\alpha\beta}\psi_\alpha(\vec{x}_1)\psi_\beta(\vec{x}_2), \tag{12.3}$$

where a_α, a_β and $a_{\alpha\beta}$ are coefficients as in a Fourier series expansion. Due to antisymmetry in the nucleon coordinates, all uncorrelated single-nucleon states ψ_m are orthogonal to each other and to ψ_{ab}. States a, b, m, ... n will be called the occupied states. For the system in the ground state, the occupied states correspond to states with momenta no greater than the Fermi momentum k_F, and they are said to be in the Fermi sea. Consequaently, in (12.3) α and β states can only be in the unoccupied states (those with momenta above k_F). In this section we use Latin indices to label occupied states and Greek indices to label unoccupied states.

Since ψ_{ab} has a definite momentum, the expression (12.3) is subject to the requirements:

$$\vec{k}_a + \vec{k}_b = \vec{k}_\alpha + \vec{k}_\beta = \vec{k}_\alpha + \vec{k}_b = \vec{k}_a + \vec{k}_\beta = \vec{K}\,. \tag{12.4}$$

Some of these conditions may be realized in a manner illustrated in Figure 12-1. On the other hand, the appearance of states (a,β) and (α,b) are forbidden since they can never satisfy (12.4). Hence, (12.3) may be simplified to read:

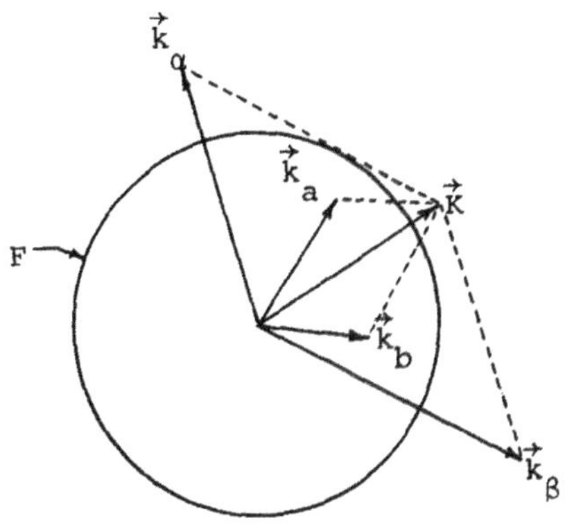

Figure 12-1. An illustration of the kinematical situation required by (12.4).

$$\psi_{ab}(\vec{x}_1,\vec{x}_2) = \psi_a(\vec{x}_1)\psi_b(\vec{x}_2) + \sum_{\alpha,\beta} a_{\alpha\beta}\psi_\alpha(\vec{x}_1)\psi_\beta(\vec{x}_2) \ . \tag{12.5}$$

For a many-body Hamiltonian in the form of:

$$H = \sum_{i=1}^{A} -\frac{\hbar^2}{2m}\nabla_i^2 + \sum_{\substack{i,j=1 \\ i<j}}^{A} v_{ij} \ , \tag{12.6}$$

where $v_{ij} = v(\vec{x}_i - \vec{x}_j)$, the many-body Schrodinger equation (12.2) is separable. The uncorrelated states obey single-nucleon equations as follows:

$$\left(-\frac{\hbar^2}{2m}\nabla^2 + U_m\right)\psi_m(\vec{x}) \equiv h_m(\vec{x})\psi_m(\vec{x}) = \varepsilon_m\psi_m(\vec{x}) \ , \tag{12.7}$$

where U_m (to be specified later) depend on the states m but are independent of the coordinates since ψ_m are plane waves. The correlated pair of nucleons feel the same overall potential as the others but are in addition experiencing their own mutual interaction as they cluster. The correlated wave function satisfies the description:

$$\left(h_a(\vec{x}) + h_b(\vec{y}) + v(\vec{x}-\vec{y})\right)\psi_{ab}(\vec{x},\vec{y}) = \varepsilon_{ab}\psi_{ab}(\vec{x},\vec{y}) \ , \tag{12.8}$$

where $v(\vec{x}-\vec{y})$ represents a realistic two nucleon potential as for example the Reid potential described in Section 11.

Inserting (12.5) into (12.8), we have:

$$(\varepsilon_a + \varepsilon_b)\psi_a\psi_b + \sum_{\alpha,\beta}(\varepsilon_\alpha + \varepsilon_\beta)a_{\alpha\beta}\psi_\alpha\psi_\beta + v\psi_{ab} = \varepsilon_{ab}\psi_{ab} \ , \tag{12.9}$$

from which the coefficient $a_{\alpha\beta}$ can be found by forming scalar products with $\psi_\alpha^+\psi_\beta^+$:

$$a_{\alpha\beta}(\varepsilon_{ab} - \varepsilon_\alpha - \varepsilon_\beta) = \int d^3x d^3y \ \psi_\alpha^+(\vec{x})\psi_\beta^+(\vec{y}) \ v(\vec{x}-\vec{y}) \ \psi_{ab}(\vec{x},\vec{y})$$

$$\equiv \langle\alpha\ \beta|\ v\ |\ \psi_{ab}\rangle \ , \tag{12.10}$$

or,

$$a_{\alpha\beta} = \frac{\langle\alpha,\beta|\ v\ |\ \psi_{ab}\rangle}{(\varepsilon_{ab} - \varepsilon_{\alpha} - \varepsilon_{\beta})} \ . \tag{12.11}$$

Thus, (12.5) gives:

$$\psi_{ab}(\vec{x},\vec{y}) = \psi_a(\vec{x})\psi_b(\vec{y}) + \sum_{\alpha,\beta} \frac{\langle\alpha,\beta|\ v\ |\ \psi_{ab}\rangle}{(\varepsilon_{ab}-\varepsilon_{\alpha}-\varepsilon_{\beta})} \psi_{\alpha}(\vec{x})\psi_{\beta}(\vec{y}) \ , \tag{12.12}$$

which is called the Brueckner-Bethe-Goldstone (BBG) equation and is often written in terms of abstract Hilbert space operators as follows:

$$|\psi_{ab}\rangle = |a,b\rangle + \frac{Q}{\varepsilon_{ab} - H_0} v\ |\psi_{ab}\rangle \ , \tag{12.13}$$

where,

$$Q = \sum_{\substack{\alpha,\beta \\ \alpha\neq\beta}} |\alpha,\beta\rangle\langle\alpha,\beta| \ , \tag{12.14}$$

is a projection operator which projects all states onto the subspace of unoccupied states (denoted by Greek indices), and:

$$H_0\ |\alpha,\beta\rangle = (\varepsilon_{\alpha} + \varepsilon_{\beta})\ |\alpha,\beta\rangle \ . \tag{12.15}$$

At this point it is convenient to introduce a quantity G, called the reaction matrix or the Brueckner G matrix, with the property that:

$$G\ |a,b\rangle = v|\psi_{ab}\rangle \ . \tag{12.16}$$

The action of G on a pair of uncorrelated wave functions is equivalent to that of v on the correlated wave function. Consequently, G has the physical interpretation of being the effective interaction employed in Chapter 2. The reaction matrix formulation is particularly suited to the nuclear matter problem where a strong repulsive core is found to exist at short range. The Hartree-Fock matrix elements of such a potential core will yield unphysically large results. In reality, the correlated wave function ψ_{ab} must be sufficient small at the core region so as to give rise to reaction matrix G which is reasonably smooth throughout.

If we apply v onto (12.13), it is converted into:

$$G\ |a,b\rangle = v\ |a,b\rangle + v \frac{Q}{\varepsilon_{ab}-H_0} G\ |a,b\rangle \ , \tag{12.17}$$

where the only reference to states (a,b) is in the energy ε_{ab}. It is therefore common to define G by the operator relation:

$$G(\omega) = v + v \frac{Q}{\omega - H_0} G(\omega) \ . \tag{12.18}$$

The BBG equation (12.17) has a similar structure to the Lippmann-Schwinger equation (11.18). In the latter case the Green's function is singular and the energy denominator is modified by an infinitesimal imaginary part to obtain an outgoing scattered wave solution. This singularity introduces a phase shift in the scattered wave. In the BBG equation, the corresponding Green's function (or the energy denominator) is not singular, since ε_{ab} (energy of states in the Fermi sea) is always less than $\varepsilon_\alpha + \varepsilon_\beta$ (energies of states above the Fermi sea), and the asymptotic form of ψ_{ab} is identical to $\psi_a\psi_b$. Only in the region where the two nucleons cluster is ψ_{ab} expected to be different from $\psi_a\psi_b$. Consequently, there is no real scattering for nucleons in nuclear matter. A novel result of the BBG equation is the "healing" of the correlated function, meaning the return of ψ_{ab} to $\psi_a\psi_b$ at long distance after a "wound" is open up as the nucleons interact at short distance.

A measure of the degree of the wound is given by the defect wave function:

$$\zeta_{ab} = \psi_a\psi_b - \psi_{ab} \ , \tag{12.19}$$

which tends to zero as the wave function heals. The wound parameter κ_{ab} is defined to be:

$$\kappa_{ab} = \int |\zeta_{ab}|^2 \, d^3x d^3y \ . \tag{12.20}$$

It represents the probability of finding the nucleon pair excited out of the Fermi sea relative to that of finding them in the occupied states (a,b). A second wound parameter characterizing the system may be defined to be:

$$\kappa = \frac{1}{A} \sum_{a<b} \kappa_{ab} \ . \tag{12.21}$$

where A is the total number of nucleons in the system. In order for the independent-pair model to be adequate in describing nuclear matter, or in other words, that the neglect of three-body or higher order correlations to be justified, κ must be small. Therefore, within the framework of the independent-pair model, κ may be consistently assumed small, since the breakdown of such an assumption would imply the breakdown of the model as well.

The scalar products of (12.9) with $\langle a,b|$ give:

$$\varepsilon_{ab} - \varepsilon_a - \varepsilon_b = \langle a,b|\ v|\ \psi_{ab}\rangle = \langle a,b|\ v\ |a,b\rangle - \langle a,b|\ v\ |\zeta_{ab}\rangle$$
$$= \Omega^{-1} \{ \int d^3r\ v(r) - \int d^3r\ e^{-i\vec{k}\cdot\vec{r}}\ v(r)\ \zeta(r) \} , \qquad (12.22)$$

where, for simplicity, spin and isospin factors have been neglected. Since both $v(r)$ and $\zeta(r)$ (which is the deflect wave function in the relative coordinates) are short-ranged, such integrals become infinitesimals after divided by a large volume Ω and may be neglected. Hence for nuclear matter it is consistent to approximate ε_{ab} by $\varepsilon_a+\varepsilon_b$.

The single-nucleon spectra from (12.7) are not observable quantities and are not related to the excitation spectrum of the system. The choice of U_m in (12.7) is therefore at our disposal. Physically, U_m should reflect the average interaction that a nucleon experiences as it is immersed in the system. It should be interaction energy between a nucleon, denoted by m, with all the other nucleons in the Fermi sea. Therefore, it is appropriate to choose:

$$U_m = \sum_n \langle m,n|\ G\ |m,n - n,m\rangle , \qquad (12.23)$$

where the summation is over all the occupied states and the ket vector on the right signifies an antisymmetrization of the uncorrelated states. Note that the difference between (12.23) and (12.22) is that in (12.23) the matrix elements are being summed over all A states and therefore even though each term is infinitesimal, as it has been claimed for (12.22), the sum is finite and significant.

For states above the Fermi sea, the single-nucleon energies may again be written as $\varepsilon_\alpha = \hbar^2 k_\alpha^2/(2m) + U_\alpha$. However, the choice of U_α is left without a clear cut perference. Practically all calculations with the BBG equation carried out before 1971 had been computed on the basis of:

$$U_\alpha = 0 , \qquad (12.24)$$

which is called the "standard choice".

The above derivation of the BBG equation is only heuristic. Historically, the independent-pair model has been treated with a much more elaborate scheme than it is described here. It starts out with Goldstone's (1957) linked cluster expansion of the many-body equation, which includes two-body as well as all higher order correlations. It is then shown by Bethe (1965) that the expansion should be resummed into what is called the "hole-line expansion". In the language of the hole-line expansion, the independent-pair model corresponds to the two-hole-line term. By a proper choice of U_m, the convergence of the expansion can be greatly improved. (12.23) turns out to be the natural choice of U_m for occupied states. It was first thought that the standard choice of (12.24) could reduce the contribution of the three-hole-line terms to be negligibly small, and thus by its employment would have included the contributions of the three-hole-line terms. This however has been shown to be a gross underestimate of the three-hole-line terms by Day (1981), who recommends the inclusion of such terms explicitly. Other choices have also been proposed (see, for instance, Lejeune and Mahaux 1978) although their justifications are only heuristic.

So far we have not prescribed a total wave function Ψ of the entire system for the independent-pair model. Ψ would be needed to evaluate the total energy of the system. Before we do this let us first examine what would be the desired expression for the total energy of the system. On heuristic ground,the total energy E in the independent-pair model should be:

$$E = \sum_a \frac{\hbar^2 k_a^2}{2m} + \frac{1}{2}\sum_{a,b} \langle a,b|\, G\, |a,b - b,a\rangle$$

$$= (0.6)\frac{\hbar^2 k_F^2}{2m} + \frac{1}{2}\sum_{a,b} \langle a,b|\, v\, |\psi_{ab} - \psi_{ba}\rangle , \qquad (12.25)$$

where the summations are over all occupied states. This result is a direct generalization of the Hartree-Fock case with the exception that the potential v is replaced everywhere by the reaction matrix G, and G is to be solved from the BBG equation. In order to define a total wave function which will give rise to this result, let us introduce the following notations. Let us write:

$$\psi_{ab} = (1 + F_{ab})\, \psi_a \psi_b , \qquad (12.26)$$

where F_{ab} take the occupied states (a,b) onto the unoccupied states (α,β) as shown in (12.5), and therefore,

$$\Psi_{ab} = (1 + F_{ab})\,\Phi_0 \,, \tag{12.27}$$

where Φ_0 is a Slater determinant of uncorrelated occupied state wave functions. The total wave function is given by:

$$\Psi = \Big(1 + \sum_{(a,b)} F_{ab} + \sum_{(a,b)}\sum_{(c,d)} F_{ab}F_{cd} + \sum_{(a,b)}\sum_{(c,d)}\sum_{(g,h)} F_{ab}F_{cd}F_{gh} + \ldots\Big)\,\Phi_0 \,, \tag{12.28}$$

where the indices a, b, c, d, ... must be distinct. For a discussion of this result see de Shalit and Weisskopf (1958), and for a more complete derivation see Kummel, Luhrmann and Zabolitzky (1978). Since F_{ab} generate unoccupied states which are orthogonal to the occupied states, we have:

$$(\Phi_0, \Psi) = 1 \,. \tag{12.29}$$

The total energy E of the system is evaluated from:

$$E = (\Phi_0, H\,\Psi) \,, \tag{12.30}$$

from which the result (12.25) is obtained.

Let us now carry out an explicit evaluation of the BBG equation. Consider first the correlated pair are in a singlet spin state, which will not be altered by the interaction and we can drop the spin wave function completely. The correlated wave function is given by:

$$\psi_{ab}(\vec{x},\vec{y}) = \frac{1}{\Omega}\, e^{i\vec{K}\cdot\vec{X}}\, \psi_{\vec{K}\vec{k}}(\vec{r}) \,, \tag{12.31}$$

where $\vec{K}$, $\vec{k}$, $\vec{X}$ and $\vec{r}$ are given by (11.2) and (11.3) with $\vec{x} = \vec{x}_a$ and $\vec{y} = \vec{x}_b$. The relative wave function ψ is indexed with $\vec{K}$ and $\vec{k}$, which are needed to specify the momentum conserving intermediate states. The uncorrelated single-nucleon wave functions for these states are combined similarly into:

$$\psi_a(\vec{x})\psi_b(\vec{y}) = \frac{1}{\Omega}\, e^{i\vec{K}\cdot\vec{X}}\, e^{i\vec{k}\cdot\vec{r}} \,. \tag{12.32}$$

After converting the summation over intermediate states into an integral over $\vec{k}'$, the BBG equation (12.12) may be rewritten as:

$$\psi_{\vec{K}\vec{k}}(\vec{r}) = e^{i\vec{k}\cdot\vec{r}} - (2\pi)^{-3}\int d^3k' \,\frac{e^{ik'\cdot r}}{e(k,k',K)}\, Q(\vec{k}',\vec{K})\int d^3r' e^{-i\vec{k}'\cdot\vec{r}'}\, v(r')\,\psi_{\vec{K}\vec{k}}(\vec{r}') \tag{12.33}$$

where,

$$e(k,k',K) = \varepsilon_\alpha + \varepsilon_\beta - \varepsilon_a - \varepsilon_b$$
$$= \frac{1}{2m}\left((\frac{\vec{K}}{2} + \vec{k}')^2 + (\frac{\vec{K}}{2} - \vec{k}')^2 - (\frac{\vec{K}}{2} + \vec{k})^2 - (\frac{\vec{K}}{2} - \vec{k})^2 \right) +$$
$$+ U_\alpha + U_\beta - U_a - U_b \,, \tag{12.34}$$

and,

$$Q(\vec{k}',\vec{K}) = \begin{cases} 1 & \text{for } |\vec{k}' \pm \vec{K}/2| > k_F \\ 0 & \text{otherwise.} \end{cases} \tag{12.35}$$

The resemblance between the BBG equation (12.33) and the scattering equation (11.18) is compelling. (12.33) can similarly be decomposed into partial waves if not for $Q(\vec{k}',\vec{K})$ which has a very complicated angular structure. In a graphical display of the momentum space, as shown in Figure 12-2, $Q(\vec{k}',\vec{K})$ includes all points outside the two spheres indicated by circles in solid lines. To simplify computation, it is a common practice to replace $Q(\vec{k}',\vec{K})$ by its angular average:

$$Q(k',K) = \frac{1}{4\pi}\int d\Omega_{\hat{k}'} Q(\vec{k}',\vec{K}) = \begin{cases} 1 & \text{if } k' > k_F + K/2 \\ (k'^2 + \frac{1}{4}K^2 - k_F^2)/k'K & \text{if } k_F + \frac{K}{2} \geq k' \geq (k_F^2 - K^2/4)^{\frac{1}{2}} \\ 0 & \text{if } k' < (k_F^2 - K^2/4)^{\frac{1}{2}} \end{cases} \tag{12.36}$$

This result can be seen from Figure 12-3. The solid circles are drawn with radii equal to k_F. The centers of these circles are separated by K. The outside dotted circle has a radius of $(k_F + K/2)$ and the inside dotted circle has a radius of $(k_F^2 - K^2/4)^{\frac{1}{2}}$. States allowed by $Q(\vec{k}',\vec{K})$ are those outside of the solid circles. Hence, all points outside of the outer dotted circles are all allowed, and thus $Q(k',K) = 1$ there. All points inside the inner dotted circle are unallowed and $Q(k',K) = 0$ there.

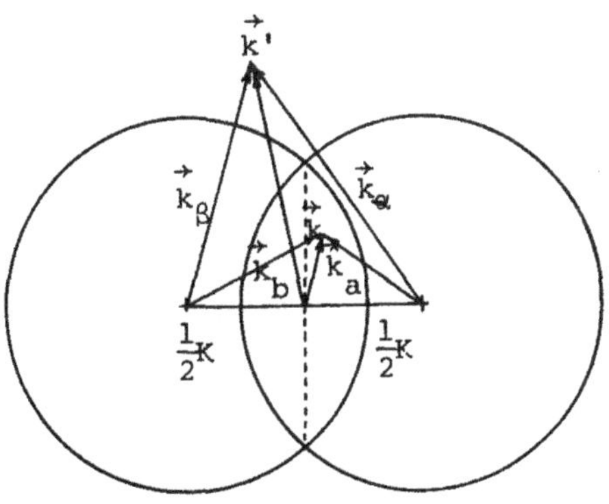

Figure 12-2. Regions in the momentum space of the occupied and unoccupied states. Momenta of the unoccupied are outside of the solid circles.

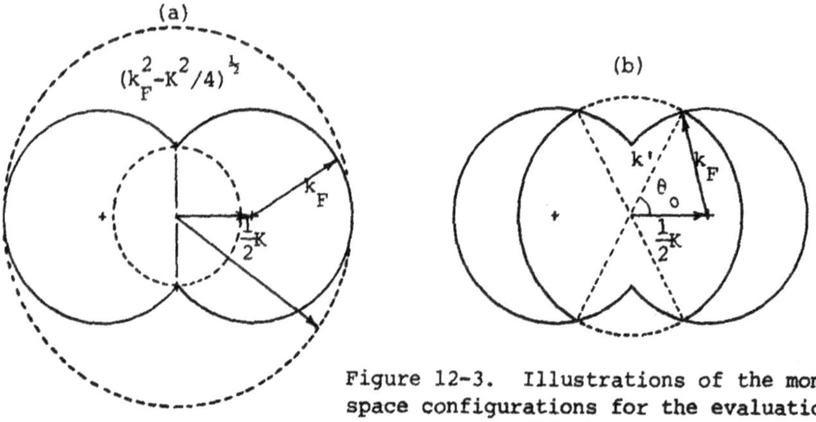

Figure 12-3. Illustrations of the momentum space configurations for the evaluations of the operator $Q(k',K)$.

The region in between requires careful evaluation. For a fixed k' only the points along the dotted arcs in Figure 12-3(b) are allowed, and hence:

$$\frac{1}{4\pi}\int d\Omega_{k'}\, Q(\vec{k}',\vec{K}) = \frac{1}{4\pi}\int_0^{2\pi} d\phi \int_{\theta_o}^{\pi-\theta_o} \sin\theta\, d\theta = \cos\theta_o \quad ,$$

where $\cos\theta_o$ can be determined from geometry as:

$$k_F^2 = k'^2 + (K/2)^2 - 2k'(K/2)\cos\theta_o \quad ,$$

or,

$$\cos\theta_o = \frac{k'^2 + (K/2)^2 - k_F^2}{k'K} \; ,$$

which is the value given by (12.36).

With $Q(\vec{k}',\vec{K})$ replaced by $Q(k',K)$ the BBG equation can be decomposed into partial waves in much the same way it is done in the case of scattering. We shall not repeat the evaluations but simply record the results. In the singlet spin case, the correlated wave function is decomposed into:

$$\psi_{\vec{K}\vec{k}} = \sum_{L=0}^{\infty} (2L+1)\, i^L\, \frac{u_L(kr)}{kr}\, P_L(\hat{k}\cdot\hat{r}) \; . \tag{12.37}$$

The partial wave BBG equations for the singlet spin state are (cf. 11.30):

$$u_L(kr) = \mathcal{J}_L(kr) - \frac{2}{\pi}\int_0^{\infty} dk'dr'\, \frac{\mathcal{J}_L(k'r)\mathcal{J}_L(k'r')}{e(k,k',K)}\, Q(k',K)\, v_\tau(r')u_L(kr') \tag{12.38}$$

where assuming charge independence we write v_τ for v_0 or v_1, the isospin singlet or isospin triplet potential, respectively. Odd-L states are associated with the isospin singlet case and even-L states with isospin triplet case. Note that the $\vec{k}'$ integration is restricted by $Q(k',K)$

For the correlated nucleons in the triplet spin states, $\psi_{\vec{K}\vec{k}}$ is decomposed as (11.56) into:

$$\psi_{\vec{K}\vec{k}} = \sum_{J=0} \sum_{L,L'=J,J\pm1} \sqrt{4\pi(2L'+1)}\, i^{L'} (Jm|L'01m) \frac{w^{L'}_{J1}}{kr} \mathcal{Y}^m_{J1L} \, . \tag{12.39}$$

The partial wave BBG equations for the triplet spin states are:

J=L:

$$w^J_{J,J} = \mathcal{J}_J(kr) - \frac{2}{\pi}\int_0^\infty dk'dr' \frac{\mathcal{J}_J(k'r)\mathcal{J}_J(k'r')}{e(k,k',K)} Q(k',K)(v^J_J)_\tau w^J_{JJ}(kr') \, ,$$

J=J±1:

$$w^{J-1}_{J,J-1} = \mathcal{J}_{J-1}(kr) - \frac{2}{\pi}\int_0^\infty dk'dr' \frac{\mathcal{J}_{J-1}(k'r)\mathcal{J}_{J-1}(k'r')}{e(k,k',K)} Q(k',K)\{(v^J_{--})_\tau w^{J-1}_{j,j-1} + (v^J_{+-})_\tau w^{J-1}_{J,J+1}\}$$

$$w^{J+1}_{J,J+1} = \mathcal{J}_{J+1}(kr) - \frac{2}{\pi}\int_0^\infty dk'dr' \frac{\mathcal{J}_{J+1}(k'r)\mathcal{J}_{J+1}(k'r')}{e(k,k',K)} Q(k',K)\{(v^J_{++})_\tau w^{J+1}_{J,J+1} + (v^J_{-+})_\tau w^{J+1}_{J,J-1}\}$$

$$w^{J+1}_{J,J-1} = -\frac{2}{\pi}\int_0^\infty dk'dr' \frac{\mathcal{J}_{J-1}(k'r)\mathcal{J}_{J-1}(k'r')}{e(k,k'K)}, Q(k',K)\{(v^J_{--})_\tau w^{J+1}_{J,J-1} + (v^J_{+-})_\tau w^{J+1}_{J,J+1}\}$$

$$w^{J-1}_{J,J+1} = -\frac{2}{\pi}\int_0^\infty dk'dr' \frac{\mathcal{J}_{J+1}(k'r)\mathcal{J}_{J+1}(k'r')}{e(k,k',K)} Q(k',K)\{(v^J_{++})_\tau w^{J-1}_{J,J+1} + (v^J_{-+})_\tau w^{J-1}_{J,J-1}\}$$

(12.40)

where v^J_J, v^J_{++}, v^J_{+-}, v^J_{-+}, v^J_{--} are given by (11.54) and the subscript τ is added to distinguish the isospin states. For triplet spin states, even-J states are associated with isospin singlet state, and odd-J states with isospin triplet states.

These equations can only be solved iteratively since the energy denominator $e(k,k'K)$ depends on U_m which according to (12.23) can only be evaluated after all the correlated wave functions are found.

Elements of the G-matrix are to be evaluated as follows. Consider first the correlated nucleons in the singlet spin state, which we indicate by adding a subscript 0 to the matrix elements:

$$\langle ab|\ G\ |ab\rangle_0 = \Omega^{-1}\int d^3r\ e^{-i\vec{k}\cdot\vec{r}}\ v(r)\ \psi_{\vec{K}\vec{k}}(r)$$
$$= \Omega^{-1}\int d^3r\Big(\sum_\ell (2\ell+1)(-i)^\ell j_\ell(kr)P_\ell(\hat{k}\cdot\hat{r})\Big)v(r)\Big(\sum_J (2J+1)i^J \frac{u_J(kr)}{kr}P_J(\vec{k}\cdot\vec{r})\Big)$$
$$= \frac{1}{\Omega}\frac{4\pi}{k^2}\sum_{J=0}^{\infty}(2J+1)\int_0^\infty dr\ J_J(kr)\ v(r)\ u_J(kr)\ . \qquad (12.41)$$

Taking isospin into consideration, the antisymmetrized matrix elements become:

$$\langle ab|\ G\ |ab - ba\rangle_0 = \frac{1}{\Omega}\frac{8\pi}{k^2}\{\sum_{J\ odd}(2J+1)\int_0^\infty dr\ J_J(kr)(v)_0\ u_J(kr) + $$
$$+ 3\sum_{J\ even}(2J+1)\int_0^\infty dr\ J_J(kr)(v)_1\ u_J(kr)\} \qquad (12.42)$$

where the subscripts of v are: 0 for isospin singlet and 1 for isospin triplet states.

For the triplet spin case (with subscript 1 on matrix elements), we write similarly:

$$\langle ab|\ G\ |ab\rangle_1 = \Omega^{-1}\sum_m\int d^3r\{\sum_{j=0}^{\infty}\sum_{\ell=j,j\pm1}\sqrt{4\pi(2\ell+1)}(-i)^\ell(jm|\ell 01m)j_\ell(kr)\ \mathcal{Y}^m_{j1\ell}\}\times$$
$$\times\{\sum_{J=0}\sum_{L,L',L''}\sqrt{4\pi(2L'+1)}\ i^{L'}(Jm|L'01m)\ \frac{w^{L'}_{J,L''}}{kr}(v)_{L''L}\ \mathcal{Y}^m_{J1L}\} \qquad (12.43)$$

Integration over $d\Omega_{\hat{r}}$ produces Kronecker deltas $\delta_{jJ}\delta_{\ell L'}$, and summation over m gives rise to factor $\delta_{LL'}(2J+1)/(2L+1)$ from (11.59), so that:

$$\langle ab|\ G\ |ab\rangle_1 = \Omega^{-1}\sum_{J,L,L''}4\pi(2J+1)\int_0^\infty r^2dr\ j_L(kr)\ \frac{w^L_{JL''}}{kr}\ (v_{L''L})_\tau$$
$$= \frac{1}{\Omega}\frac{4}{k^2}\sum_{J=0}^{\infty}(2J+1)\int_0^\infty dr\{\ J_J(kr)(v_J)_\tau w^J_{JJ} + J_{J-1}(kr)\big((v_{--})_\tau w^{J-1}_{J,J-1}+(v_{+-})_\tau w^{J-1}_{J,J+1}\big)+$$
$$+ J_{J+1}(kr)\big((v_{++})_\tau w^{J+1}_{J,J+1}+(v_{-+})_\tau w^{J+1}_{J,J-1}\big)\} \qquad (12.44)$$

Finally, the reaction matrix elements for correlated nucleon pairs (in all spin and isospin configurations) with center-of-mass momentum $\vec{K}$ and relative momentum $\vec{k}$ are given by:

$$\langle ab|\ G\ |ab - ba\rangle = \langle ab|\ G\ |ab - ba\rangle_0 + \langle ab|\ G\ |ab - ba\rangle_1$$

$$= \frac{8\pi}{\Omega k^2} \sum_{J\ \text{odd}} (2J+1) \int_0^\infty dr \{ \mathcal{J}_J\, v_0\, u_J + \mathcal{J}_{J-1}\left((v_{--})_0 w^{J-1}_{J,J-1} + (v_{+-})_0 w^{J-1}_{J,J+1} \right) +$$

$$+ \mathcal{J}_{J+1}\left((v_{++})_0 w^{J+1}_{J,J+1} + (v_{-+})_0 w^{J+1}_{J,J-1} \right) + 3\, \mathcal{J}_J (v_J)_1 w^J_{J,J} \} +$$

$$+ \frac{8\pi}{\Omega k^2} \sum_{J\ \text{even}} (2J+1) \int_0^\infty dr \{ 3\, \mathcal{J}_J\, v_1\, u_J + 3\, \mathcal{J}_{J+1}\left((v_{--})_1 w^{J-1}_{J,J-1} + (v_{+-})_1 w^{J-1}_{J,J+1} \right) +$$

$$+ 3\, \mathcal{J}_{J+1}\left((v_{++})_1 w^{J+1}_{J,J+1} + (v_{-+})_1 w^{J-1}_{J,J-1} \right) + \mathcal{J}_J (v_J)_0 w^J_{J,J} \} .$$

(12.45)

To evaluate the single-nucleon potential U_m we need to sum up the effective interaction of nucleon m with all the other nucleons in the Fermi sea as given by (12.33). Since the reaction matrix elements are expressed in terms of the relative momentum $\vec{k}$, we need an expression converting the summation over states into an integral over a density of states in the k-space:

$$\sum_n \rightarrow \int n(k,k_m)\, dk \ , \qquad (12.46)$$

where $n(k,k_m)$ can be found as follows. Referring to Figure 12-4, all $\vec{k}_n$ states to be summed fall within the large sphere of radius k_F, i.e., $|\vec{k}_n| \le k_F$. Starting with any $\vec{k}_m$, all states having a relative momentum $\vec{k}$ with nucleon m fall between the double spheres of radius 2k and width d(2k). The states within the double spheres are given by the following solid angle measurement:

$$\text{solid angle} = 2\pi \int_{\theta_1}^{\pi} \sin\theta\, d\theta = 2\pi\,(1 + \cos\theta_1) \ ,$$

where θ_1 is given implicitly by: $k_F^2 = k_m^2 + 4k^2 + 4kk_m\cos\theta_1$. The density

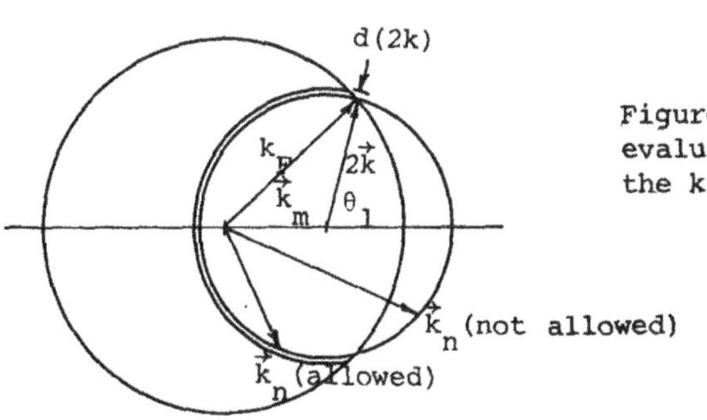

Figure 12-4. Configuration for the evaluabtion of density of states in the k-space.

states is therefore:

$$n(k,k_m)dk = \gamma\Omega(2\pi)^{-2}(1 + \cos\theta_1)(2k)^2\, d(2k) .$$

However, we must be careful to note that when $(k_m+2k) \leq k_F$, then $\cos\theta_1=1$. Combining these two situations, we have:

$$n(k,k_m)dk = \gamma\Omega(2/\pi^2)k^2dk\,\{1 + \eta(k,k_m)\} = 12A\,\frac{k^2dk}{k_F^3}\{1 + \eta(k,k_m)\} ,$$

where

$$\eta(k,k_m) = \begin{cases} 1 & \text{for } k \leq \frac{1}{2}(k_F-k_m) \\ (k_F^2-k_m^2-4k^2)/(4kk_m) & \text{for } \frac{1}{2}(k_F-k_m) < k \leq \frac{1}{2}(k_F+k_m) . \end{cases} \qquad (12.47)$$

Thus,

$$U_m = \int_0^{\frac{1}{2}(k_F+k_m)} n(k,k_m)dk\, \langle m,n|\, G\, |m,n - n,m\rangle . \qquad (12.48)$$

These are the single-nucleon potentials which go into the evaluation of the energy denominator $e(k,k',K)$ of the partial wave BBG equations for the determination of the radial wave functions u and w. The computation is finally accomplished when a self-consistent set of U_m for all occupied states is found.

The total energy of the system expressed as energy per nucleon is given by:

$$\frac{E}{A} = (0.6)\frac{\hbar^2}{2m}k_F^2 + \frac{1}{2A}\sum_n U_n , \qquad (12.49)$$

for which the summation over occupied states can be replaced by:

$$\sum_n \rightarrow \gamma\Omega(2\pi)^{-3}\int_0^{k_F} 4\pi\, k_n^2\, dk_n = 3A(k_F)^{-3}\int_0^{k_F} k_n^2\, dk_n . \qquad (12.50)$$

Nuclear matter studies by means of the BBG equation have been reviewed exhaustively in articles by Bethe (1971), Day (1967, 1978), and Svenne (1979). We refer the interested readers to these articles for details. Nuclear matter is shown to saturate. The computed saturation energy and density, however, do not agree well with empirical results. The situation is shown in Figure 12-5, which plots the average energy per nucleon at saturation against the saturation density (expressed in the Fermi momentum k_F). Each point corresponds to the results computed from a specific set of phenomenological two-nucleon potentials. The empirical saturation point lies in the open rectangle.

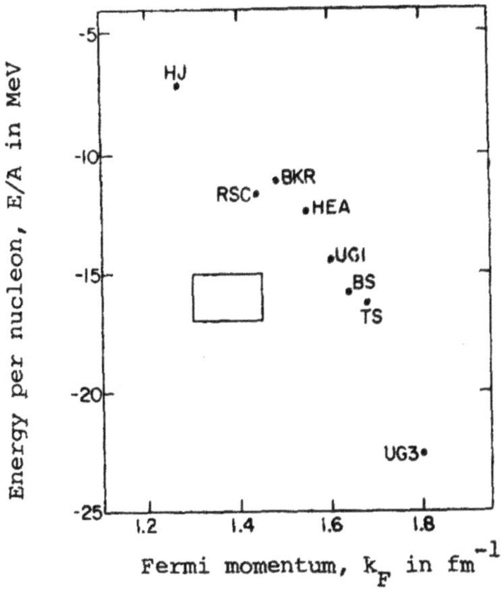

Figure 12-5. Nuclear matter saturation energy and density (given by Fermi momentum k_F) computed with the Brueckner-Bethe-Goldstone method with various potentials. See Day (1978) for potential designations. The empirical saturation lies in the open rectangle.

Figure 12-5 is taken from Day (1978) and references to the data are given there. The point marked RSC corresponds to results computed from the Reid potentials given in Section 11. All computed saturation points fall approximately on a band called the Coester band. Such a band is the result of an inherent ambiguity in the nuclear potentials which are constructed from scattering phase shifts. Even though all potentials will reproduce the same set of phase shifts, they nevertheless predict saturation properties falling on different parts of the band.

The general feeling towards the BBG approach is that even though the two-body correlation part, which it treats accurately, gives the bulk of the binding energy to nuclear matter, the other neglected parts are not entirely negligible. We list in Table 12-1 the compilation of results by Bethe (1972) on the estimated contributions of other sources to the nuclear matter energy, E/A, at three different densities. The two-body correlations term refers to energies computed from the BBG equation. The others have not been discussed here (see Bethe 1971). Experimental binding energy per nucleon is estimated to be 15.68 MeV at $k_F = 1.36\ fm^{-1}$, or $\rho = 2.7 \times 10^{14}\ g/cm^3$. It is remarkably close to the saturation energy given in Table 12-1 at that density. More recent calculations by Day (1981) put the contribution of the three-body correlations term to be -5 MeV instead of -1.76 MeV as shown.

Table 12-1. Contributions to nuclear matter energy

Contribution	$k_F = 1.2\ fm^{-1}$ $\rho = 1.85x10^{14}$	$k_F = 1.36\ fm^{-1}$ $\rho = 2.7x10^{14}$	$k_F = 1.6\ fm^{-1}$ $\rho = 4.6x10^{14}\ g/cm^3$
2-body correlations	-9.79 MeV	-11.05 MeV	-10.20 MeV
(3-body correlations)	-1.6	-1.76	-2.6
(4-body correlations)	-0.9	-1.09	-1.5
(3-body forces)	-0.8	-1.0	-1.0
minimal relativity	-0.35	-0.5	-0.7
Total	-13.3	-15.4	-16.0

The program is not yet in a completely satisfactory condition.

Of particular interest to the study of the structure of the neutron star is the equation of state for pure neutron matter at approximately the nuclear matter density. The method just described if adequate for nuclear matter should be adequate for neutron matter as well. Study of neutron matter with the BBG equation has been carried out by Siemens and Pandharipande (1971), Wang, Rose and Schlenkar (1970), (see also Leung and Wang 1971). Energy per neutron, E/A, may be computed from expressions for nuclear matter by retaining just the two-nucleon total isospin T=1 terms, dropping the (2T+1)=3 factors and taking the degeneracy factor to be 2. The results of Siemens and Pandharipande are shown in Table 12-2, in which the interaction as well as the kinetic

Table 12-2. Binding energy per particle (MeV) of neutron matter at various k_F.

$k_F(fm^{-1})$	0.3	0.5	0.75	1.0	1.4	1.7	2.1	2.5
Total interaction	-0.34	-1.25	-3.29	-6.51	-13.92	-19.9	-24.5	-22.3
Kinetic energy	1.12	3.11	7.00	12.44	24.38	36.0	54.9	77.8
E/A	0.78	1.86	3.71	5.93	10.46	16.0	30.4	55.4

energy contributions are listed.

The equation of state for pure neutron matter is shwon in Figure 12-6, which is taken from Wang, Rose and Schlenkar computed with the Reid potentials. The free neutron curve is also plotted for comparison.

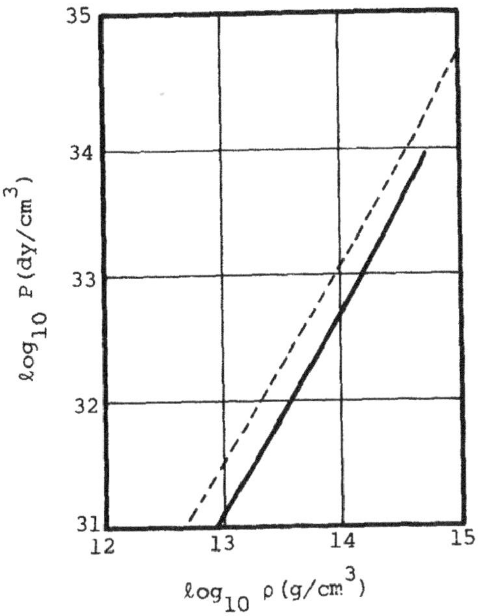

Figure 12-6. The equation of state for pure neutron matter computed by the Brueckner-Bethe-Goldstone method with the Reid potentials by Wang, Rose and Schlenkar. The equation of state of a free neutron gas is plotted in dotted lines.

There are two general approximate approaches to the solution of the many-body problem. One is the perturbative approach and the other is the variational approach. The method of Brueckner, Bethe and Goldstone discussed here belongs to the perturbative approach. It is the first order result from a perturbative expansion of the many-body theory. Estimates of its departure from the exact results are made by computing the higher order perturbative terms. To this accuracy the variational method is equally viable in nuclear physics, and constitutes an alternative approach to the problem. In the next section, we shall describe a variational method which has gained popularity among nuclear physicists lately.

References

Bethe, H.A. (1956). Phys. Rev. 103, 1353.

Bethe, H.A. (1965). Phys. Rev. B138, 804.

Bethe, H.A. (1971). Ann. Rev. Nucl. Sci. 21, 93.

Brueckner, K.A. (1954). Phys. Rev. 96, 508.

Day, B.A. (1967). Rev. Mod. Phys. 39, 719.

Day, B.A. (1978). Rev. Mod. Phys. 50, 495.

Day, B.A. (1981). Phys. Rev. C24, 1203.

De-Shalit, A. and Weisskopf, V.F. (1958). Ann. Phys. 5, 282.

Goldstone, J. (1957). Proc. Roy. Soc. (London) A239, 267.

Kummel, H., Luhrmann, L.H. and Zabolitzky, J.G. (1978). Phys. Rep. 36, 1.

Lejeune, A. and Mahaux, C. (1978). Nucl. Phys. A295, 169.

Leung, Y.C. and Wang, C.G. (1971). Astrophys. Jl. 170, 499.

Siemens, P.J. and Pandharipande, V.R. (1971). Nucl. Phys. A173, 561.

Svenne, J.P. (1979). In Advances in Nuclear Physics, (Ed. J.W. Negele and E. Vogt), Vol. 11, p.179. Plenum Press, New York.

Wang, C.G., Rose, W.R. and Schlenker, S.L. (1970). Astrophys. Jl. 160, L17.

Bibliography

Brueckner, K.A. (1959). In The Many-Body Problems, (Ed. C. DeWitt). Dunod Cie., Paris.

Coester, F. (1969). In Lectures in Theoretical Physics, (Ed. K.T. Mahanthappa and Brittin, W.E.), Vol. 11B. Gordon and Breach, New York.

Gomes, L.C., Walecka, J.D. and Weisskopf, V.F. (1958). Ann. Phys. 3, 241.

Preston, M.A. and Bhaduri, R.K. (1975). Structure of the Nucleus, Addison-Wesley, Reading, Mass.

13. The Constrained Variational Method

The variational approach to the nuclear independent-pair model starts with a trial wave function called the Jastrow trial wave function written in the following form:

$$\Psi = \sum_{i<j} f(ij)\, \Psi_0 \equiv F\, \Psi_0 , \qquad (13.1)$$

where Ψ_0 represents the Hartree-Fock wave function (1.17) and f(ij) are the two-nucleon correlation factors for nucleons in states i and j (Jastrow 1955). The correlation factors f(ij) are assumed to be independent of states i and j and would have the same functional form for all (i,j) pairs. F is therefore symmetric while Ψ_0 is antisymmetric in the particle coordinates. However, the correlation factors would vary with the state of the total system, so that f(ij) would vary, for example, with the density of the system. In the original application of (13.1) by Jastrow, f(ij) are further assumed to be dependent on the interparticle separations $r_{ij} = |\vec{x}_i - \vec{x}_j|$ only. In the more recent applications, f(ij) incorporate spin and isospin dependent parts. f(ij) will be taken to be real-valued functions, and have the value unity whenever the interparticle separations becomes large:

$$f(ij) \to 1 , \qquad \text{for } r_{ij} > d , \qquad (13.2)$$

where d is a length parameter determined by the nature of the interaction. Two nucleons are said to be clustering when their separation is within d.

The variational principle gives the ground state energy to be:

$$E = \text{Min} \left\{ \frac{\langle\Psi|\, H\, |\Psi\rangle}{\langle\Psi|\Psi\rangle} \right\} , \qquad (13.3)$$

where the Hamiltonian H is given by (11.1). The minimization requirement will impose conditions on the correlation factors f(ij). We shall consider the nuclear matter problem. Due to uniformity of the system, the single-nucleon wave functions in Ψ_0 are given by plane waves. Let us isolate a finite volume Ω within the infinitely extended system. There will be a total of A nucleons in Ω, and A will be large. The expectation value of the Hamiltonian is given by:

$$\bar{H} = \frac{\langle\Psi|H|\Psi\rangle}{\langle\Psi|\Psi\rangle} =$$

$$= \int d^3x_1 \cdot \cdot \cdot d^3x_A \; \Psi_0^* F \{-\frac{\hbar^2}{2m}\sum_{i=1} \nabla_i^2 + \sum_{i<j} v_{ij}\} F\Psi_0 \; / \; \langle\Psi|\Psi\rangle \; . \quad (13.4)$$

(13.4) as it stands is still in an untractable form. A systematic evaluation of $\bar{H}$ relies on an expansion called the cluster seris expansion. The expansion parameter corresponds to the products of the nucleon number density n_B of the system with the volume integral of $h(ij) = f(ij)^2 - 1$. Hence this expansion is of practical value if the average interparticle separation is large compared to the range d of the correlation factors. In the following, we shall assume this to be the case and discuss a method which keeps just the lowest order term in such an expansion. We shall not develop the cluster series expansion here which has been presented in several review articles (see, for example, Clark 1979), but merely indicate how the lowest order term may be obtained. The variational solution of $\bar{H}$ in the lowest order approximation leads however to correlation factors which are not consistent with (13.2). This is because the long-ranged attractive forces between two nucleons cause the approximate $\bar{H}$ to be lowered without limit by favoring long-ranged correlated f(ij). This catastrophic result can be avoided if higher order cluster series terms are kept, but they are not. In order to make the lowest order variational method to produce results, an ingenious idea of incorporating physically acceptable constraints on f(ij) was proposed (Pandharipande 1971b). This method works unexpectedly well, and will be the topic of discussion in this section. It is called the Lowest Order Constrained Variational (LOCV) method.

Let us first evaluate the interaction term. From the symmetry of the wave function it may be reduced to read:

$$\bar{V} = \frac{1}{2} A(A-1) \int d^3x_1 d^3x_2 f(12) v_{12} f(12) \left(\Omega^{-2} G(12)\right) , \quad (13.5)$$

where,

$$G(12) = \frac{\Omega^2 \int d^3x_3 . . . d^3x_A \; \Psi_0^* \left(F^2/f(12)^2\right) \Psi_0}{\int d^3x_1 d^3x_2 d^3x_3 . . . d^3x_A \; \Psi_0^* F^2 \Psi_0} \; . \quad (13.6)$$

If we ignore the plane wave factors containing in Ψ_0, then to the lowest order in the cluster series expansion $G(12)$ is given by unity. The first order expansion term in $G(12)$ is shown by Jastrow to be $n_B \int d^3x_3 h(13)h(23)$, which we shall drop. In the present case, to the lowest order:

$$G(12) = \frac{1}{A(A-1)} \sum_{i,j} \psi_i^*(\vec{x}_1)\psi_j^*(\vec{x}_2) \{\psi_i(\vec{x}_1)\psi_j(\vec{x}_2) - \psi_j(\vec{x}_1)\psi_i(\vec{x}_2)\} \tag{13.7}$$

where ψ_i are the plane wave single-nucleon states forming Ψ_0 (see 7.26). The interaction term is given by:

$$\bar{V} = \Omega^{-2} \frac{1}{2} \sum_{i,j} \langle i,j| f(12)v_{12}f(12) |i,j - j,i\rangle , \tag{13.8}$$

in the notations of Section 12.

The kinetic energy part is reduced similarly to read:

$$\bar{T} = T_1 + At_F , \tag{13.9}$$

where,

$$T_1 = - A \frac{\hbar^2}{2m} \int d^3x_1 \ldots d^3x_A \Psi_0^* \{F(\nabla_1^2 F)+2F(\vec{\nabla}_1 F)\cdot\vec{\nabla}\}\Psi_0 / \langle\Psi|\Psi\rangle , \tag{13.10}$$

and,

$$t_F = - \frac{\hbar^2}{2m} \int d^3x_1 \ldots d^3x_A \Psi_0^* F^2(\nabla^2\Psi_0) / \langle\Psi|\Psi\rangle = (0.6)\frac{\hbar^2}{2m} k_F^2 . \tag{13.11}$$

The gradient operator acts on F as follows:

$$\vec{\nabla}_1 F = (\sum_k \frac{\vec{\nabla}_1 f(1k)}{f(1k)})F = (A-1)\frac{\vec{\nabla}_1 f(12)}{f(12)} F, \tag{13.12}$$

and,

$$\nabla_1^2 F = (A-1)(\frac{\nabla_1^2 f(12)}{f(12)})F + \sum_{j\neq k\neq 1} \frac{\vec{\nabla}_1 f(1j)}{f(1j)} \cdot \frac{\vec{\nabla}_1 f(1k)}{f(1k)} F. \tag{13.13}$$

The crossed-terms in $\nabla_1^2 F$, however, do not contribute to T_1, since upon integration over $\vec{x}_k$, $f(1k)$ will no longer be dependent on $\vec{x}_1$ and $\vec{\nabla}_1 f(1k)$ will vanish. Hence, T_1 in the lowest order approximation is given by:

$$\begin{aligned} T_1 &= A(A-1)\Omega^{-2}\int d^3x_1 d^3x_2 \psi_i^*(\vec{x}_1)\psi_j^*(\vec{x}_2)(-\frac{\hbar^2}{2m})\{f(12)\nabla_1^2 f(12)+2f(12)\vec{\nabla}_1 f(12)\cdot\vec{\nabla}_1\}\times \\ &\qquad \times\{\psi_i(\vec{x}_1)\psi_j(\vec{x}_2) - \psi_j(\vec{x}_1)\psi_i(\vec{x}_2)\} \\ &= \Omega^{-2}\sum_{i,j} \langle i,j| - \frac{\hbar^2}{2m}\{f(12)\nabla_{12}^2 f(12)+2f(12)\vec{\nabla}_{12} f(12)\cdot\vec{\nabla}_{12}\} |i,j - j,i\rangle , \end{aligned} \tag{13.14}$$

where $\vec{\nabla}_{12}$ refers to the gradient in the relative coordinates $\vec{r}=\vec{x}_1-\vec{x}_2$.

With the form of the correlation factors which we shall adopt some ambiguity in the kinetic energy term is introduced. The fact is that $\vec{\nabla}f$ and f do not commute, and the kinetic energy is evaluated with different results depending on whether it is written as $\Psi_0^* F(\nabla^2 F\Psi_0)$ or $(\nabla^2\Psi_0^* F)F\Psi_0$. In what follows we shall adopt a compromised estimate of the kinetic energy by evaluating it from:

$$\frac{1}{2}\{\Psi_0^* F\ (\nabla_1^2 F\Psi_0) + (\nabla_1^2\Psi_0^* F)\ F\Psi_0\} = \frac{1}{2}\Big(\Psi_0^* F^2(\nabla_1^2\Psi_0) + (\nabla_1^2\Psi_0^*)F^2\Psi_0 +$$

$$+\Psi_0^*\{F(\nabla_1^2 F)+(\nabla_1^2 F)F\}\Psi_0 + 2(\vec{\nabla}_1\Psi_0^*\cdot\vec{\nabla}_1 F)F\Psi_0 + 2(\Psi_0^* F)(\vec{\nabla}_1 F\cdot\vec{\nabla}_1\Psi_0)\Big). \qquad (13.15)$$

Different forms of the kinetic energy term have been introduced because they have different cluster series expansions, and some are more convenient to evaluate than the others. In the present case, computing $\overline{H}$ to the lowest order, the choice of (13.15) is also based on convenience. It is clear that the first two terms on the right of (13.15) contribute to the Fermi energy (13.11), while the remaining terms, after performing an integration by parts to remove the gradient on $\vec{\nabla}\Psi_0^*$, can be expressed compactly as follows:

$$\Psi_0^*[\,F,[\nabla_1^2,\ F]]\Psi_0 = \Psi_0^*\ \{F(\nabla_1^2 F) - (\nabla_1^2 F)F + 2F(\vec{\nabla}_1 F)\cdot\vec{\nabla}_1 +$$

$$- 2(\vec{\nabla}_1 F)F\cdot\vec{\nabla}_1 - 2(\vec{\nabla}_1 F)(\vec{\nabla}_1 F)\}\ \Psi_o\ . \qquad (13.16)$$

With kinetic energy given in this form and to the lowest order approximation, T_1 is reduced to:

$$T_1 = \frac{1}{2}\sum_{i,j} \langle i,j| - \frac{\hbar^2}{2m}[f(12),[\nabla_{12}^2,\ f(12)]]\ |i,j - j,i\rangle\ . \qquad (13.17)$$

In (13.17) we have replaced ∇_1^2 by $\frac{1}{2}(\nabla_1^2+\nabla_2^2) = (\nabla_{12}^2+\frac{1}{4}\nabla_C^2)$, where $\vec{\nabla}_C$ denotes the gradient in the center-of-mass coordinates. Terms related to the center-of-mass motion will eventually be cancelled out and will be ignored. f(12) and v_{12} are functions of the two-nucleon relative coordinates and will be written simply as f and v, respectively, while $\vec{\nabla}_{12}$ as $\vec{\nabla}$. In summary, the expectation value of H is given by:

$$\overline{H} = (0.6)\frac{\hbar^2}{2m}k_F^2 A + \frac{1}{2}\sum_{i,j}\langle i,j| - \frac{\hbar^2}{2m}[f,[\nabla^2,f]] + fvf|i,j - j,i\rangle\ . \qquad (13.18)$$

We shall next decompose the two-nucleon states into partial waves with the intention of employing the Reid potentials for v(r), which is expressed as;

$$v(r) = v_C + v_T S_{12} + v_{LS}\vec{L}\cdot\vec{S} , \tag{13.19}$$

with the understanding that v_C, v_T, and v_{LS} are different for different partial waves. The uncorrelated two-nucleon states are given by:

$$|i,j\rangle = \Omega^{-1} e^{i\vec{k}_i\cdot\vec{x}} e^{i\vec{k}_j\cdot\vec{y}} \chi(i)\chi(j) = \Omega^{-1} e^{i\vec{K}\cdot\vec{R}} e^{i\vec{k}\cdot\vec{r}} \chi_S^m =$$

$$= \phi_C \Omega^{-\frac{1}{2}} \sum_{J=0}^{\infty} \sum_{L=J,J\pm 1} \sqrt{4\pi(2L+1)}\, i^L (Jm|L0Sm) j_L(kr) \mathcal{Y}^m_{JSL}(\hat{k}\cdot\hat{r}), \tag{13.20}$$

where $\phi_C = \Omega^{-\frac{1}{2}} e^{i\vec{K}\cdot\vec{R}}$ denotes the center-of-mass wave function which we shall drop, while the partial wave expansion follows from (11.55). Among all the (JLS) states those which do not couple under potential (13.19) are: (1) S=0, J=L, and (2) S=1, J=L, and those which form coupled states are labelled S=1, J=L±1.

The correlation factors f(12) will have components in each of these partial wave channels, and are decomposed by means of projection operators as follows:

$$f(12) = \sum_J \left\{ f_J^{(0)} P_J^0 + f_J^{(1)} P^1_{J=L} + \left(f_J^{(2)} P_J^{(2)} + f_J^{(3)} P_J^{(3)}\right) P^1_{J=L\pm 1}\right\} , \tag{13.21}$$

where the P's are projection operators projecting from the two-nucleon state partial waves of designated (JLS): P_J^0 for S=0, J=L; $P^1_{J=L}$ for S=1, J=L, and $P^1_{J=L\pm 1}$ for S=1, J=L±1. $P_J^{(2)}$ and $P_J^{(3)}$ are additional projection operators within the set of J=L±1 states. They take on the form:

$$P_J^{(i)} = a_i + b_i S_{12} + c_i \vec{L}\cdot\vec{S} , \quad (i = 2,3) \tag{13.22}$$

subject to the requirements that:

$$P_J^{(i)} P_J^{(j)} = \delta_{ij} P_J^{(i)} . \tag{13.23}$$

Strictly speaking the coefficients a_i, b_i, c_i of (13.22) should be treated as variational parameters to yield the best set of projection operators. But as suggested by Ristig, TerLouw and Clark (1971), to simplify matter one may assume the spin-orbit effects in a uniform system to be small (compared to the tensor effects) and take $c_i = 0$.

$P_J^{(2)}$ and $P_J^{(3)}$ satisfying (13.23) may then be chosen to be:

$$P_J^{(2)} = \frac{2}{3} + \frac{1}{6} S_{12} ,$$
$$P_J^{(3)} = \frac{1}{3} - \frac{1}{6} S_{12} , \tag{13.24}$$

which are independent of J and the subscript J will be dropped. Hence, the correlated two-nucleon states are decomposed into:

$$|\psi_{ij}\rangle = f(12)|i,j\rangle = \Omega^{-\frac{1}{2}}\{\sum_{J=0}^{\infty} \sqrt{4\pi(2J+1)}\, i^J j_J(kr)\left(f_J^{(0)} Y_{J0}\chi_0 + f_J^{(1)} \mathcal{Y}_{J1J}^m\right) +$$
$$+ \sum_{J=0}^{\infty} \sum_{L=J\pm1} \sqrt{4\pi(2L+1)}\, i^L (Jm|L0Sm) j_L(kr)\left(f_J^{(2)} P^{(2)} + f_J^{(3)} P^{(3)}\right) \mathcal{Y}_{J1L}^m\} \tag{13.25}$$

where Y_{J0} are the spherical harmonics, χ_0 is the singlet spin wave function, and $\mathcal{Y}_{JSL}$ are the vector spherical harmonics given by (11.38).

The matrix elements of the potential, denoted by:

$$M_V(ij) = \langle\Psi_{ij}|\, v_C + v_T S_{12} + v_{LS}\vec{L}\cdot\vec{S}\, |\Psi_{ij}\rangle \equiv M_V^{(uncoupled)} + M_V^{(coupled)} \tag{13.26}$$

will consist of parts which show different degrees of complexity depending on the type of partial waves. The part involving uncoupled waves, S = 0 or 1, J = L, are relatively simple and is given by:

$$M_V^{(uncoupled)} = \frac{1}{\Omega} \sum_J 4\pi(2J+1)\int_0^{\infty} r^2 dr\left(j_J(kr)\right)^2 \{v_C\left(f_J^{(0)}\right)^2 + (v_C+2v_T-v_{LS})\left(f_J^{(1)}\right)^2\}. \tag{13.27}$$

With the Reid potentials, v_C and $(v_C+2v_T-v_{LS})$ will be replaced by specific potentials appropriate for each J state. As we can see, the dependence of $M_V(ij)$ on states (i,j) is entirely through their relative momentum $\vec{k} = \vec{k}_i - \vec{k}_j$.

The matrix elements of the potential involving the coupled states are given by:

$$M_V^{(coupled)} = \frac{1}{\Omega} \sum_{J,L,L'} 4\pi\sqrt{(2L+1)(2L'+1)}\; i^{(L-L')}\, j_L(kr) j_{L'}(kr) \times$$
$$\times \sum_{m=-1}^{1} (Jm|L0Sm)(Jm|L'0Sm) \int d\Omega\, \mathcal{Y}_{J1L'}^{m=0}\left(f_J^{(2)} P^{(2)} + f_J^{(3)} P^{(3)}\right) \times$$
$$\times\; v(r)\left(f_J^{(2)} P^{(2)} + f_J^{(3)} P^{(3)}\right) \mathcal{Y}_{J1L}^{m=0} =$$

$$= \frac{1}{\Omega} \sum_J \sum_{L=J\pm 1} 4\pi(2J+1) \int r^2 dr \; j_L^2(kr) \int d\Omega \; \mathcal{Y}^0_{J1L} \left(f_J^{(2)} P^{(2)} + f_J^{(3)} P^{(3)} \right) \times$$

$$\times \; v(r) \; \left(f_J^{(2)} P^{(2)} + f_J^{(3)} P^{(3)} \right) \mathcal{Y}^0_{J1L} \quad , \qquad (13.28)$$

where in the second line, m is set equal to 0 for all $\mathcal{Y}_{J1L}$ since the result will be independent of m, and the summation over m for the C-G coefficients is evaluated according to (11.59). To evaluate (13.28) further, let us first note the following results:

$$(2J+1)\; P^{(2)} \; \mathcal{Y}^m_{J1J-1} \;=\; (J+1)\; \mathcal{Y}^m_{J1J-1} + \sqrt{J(J+1)}\; \mathcal{Y}^m_{J1J+1} \; ,$$

$$(2J+1)\; P^{(2)} \; \mathcal{Y}^m_{J1J+1} \;=\; J\; \mathcal{Y}^m_{J1J+1} + \sqrt{J(J+1)}\; \mathcal{Y}^m_{J1J-1} \; ,$$

$$(2J+1)\; P^{(3)} \; \mathcal{Y}^m_{J1J-1} \;=\; J\; \mathcal{Y}^m_{J1J-1} + \sqrt{J(J+1)}\; \mathcal{Y}^m_{J1J+1} \; ,$$

$$(2J+1)\; P^{(3)} \; \mathcal{Y}^m_{J1J+1} \;=\; (J+1)\; \mathcal{Y}^m_{J1J+1} + \sqrt{J(J+1)}\; \mathcal{Y}^m_{J1J-1} \; , \qquad (13.29)$$

which follow from (11.52). Next, we shall try to commute all $P^{(i)}$ projectors over to the right of v(r). In so doing we need the results that:

$$P^{(2)} (\vec{L}\cdot\vec{S}) \;=\; (\vec{L}\cdot\vec{S}) P^{(3)} + 2P_3 - 3P^{(2)} \; ,$$

$$P^{(3)} (\vec{L}\cdot\vec{S}) \;=\; (\vec{L}\cdot\vec{S}) P^{(2)} - 2P_3 + 3P^{(2)} \; , \qquad (13.30)$$

where,

$$P_3 \;=\; \frac{1}{4}(\vec{\sigma}_1\cdot\vec{\sigma}_2 + 3) \; , \qquad (13.31)$$

is the projector for the triplet spin states. These and other useful results are given by Ristig, Ter Louw and Clark (1971). (13.31) may be found from a direct evaluation as follows. It is convenient to express $P^{(2)}$ as:

$$P^{(2)} \;=\; (\vec{S}\cdot\hat{r})^2 \;=\; \frac{1}{2}\left((\vec{\sigma}_1\cdot\hat{r})(\vec{\sigma}_2\cdot\hat{r}) + 1 \right) \;=\; 1 - P^{(3)} .$$

Then,

$$(\vec{L}\cdot\vec{S})(\vec{S}\cdot\hat{r})^2 \;=\; \frac{1}{4}(\vec{\sigma}_1+\vec{\sigma}_2)\cdot\vec{r}\times\vec{p}(\vec{\sigma}_1\cdot\hat{r})(\vec{\sigma}_2\cdot\hat{r}) + \frac{1}{2}\vec{L}\cdot\vec{S} \; ,$$

where $\vec{p} = (\hbar/i)\vec{\nabla}$. Pushing $\vec{p}$ through the expression and making use of the property that $\vec{\sigma}\times\vec{\sigma} = 2i\vec{\sigma}$ we have:

$$\vec{\sigma}_1\cdot\vec{r}\times\vec{p}(\vec{\sigma}_1\cdot\hat{r})(\vec{\sigma}_2\cdot\hat{r}) \;=\; -3(\vec{\sigma}_1\cdot\hat{r})(\vec{\sigma}_2\cdot\hat{r}) + (\vec{\sigma}_1\cdot\vec{\sigma}_2) + ir(\vec{\sigma}_2\cdot\hat{r})(\vec{\sigma}_1\cdot\vec{p}) +$$

$$-i(\vec{\sigma}_1\cdot\hat{r})(\vec{\sigma}_2\cdot\hat{r})\vec{r}\cdot\vec{p} \; ,$$

and since,

$$(\vec{\sigma}_1\cdot\hat{r})(\vec{\sigma}_2\cdot\hat{r})\vec{\sigma}_1\cdot\vec{r}\times\vec{p} = -ir(\vec{\sigma}_2\cdot\hat{r})(\vec{\sigma}_1\cdot\vec{p}) + i(\vec{\sigma}_1\cdot\hat{r})(\vec{\sigma}_2\cdot\hat{r})\vec{r}\cdot\vec{p} ,$$

it follows that,

$$(\vec{L}\cdot\vec{S})(\vec{S}\cdot\hat{r})^2 + (\vec{S}\cdot\hat{r})^2(\vec{L}\cdot\vec{S}) = -\frac{3}{2}(\vec{\sigma}_1\cdot\hat{r})(\vec{\sigma}_2\cdot\hat{r}) + \frac{1}{2}(\vec{\sigma}_1\cdot\vec{\sigma}_2) + \vec{L}\cdot\vec{S} ,$$

from which the results of (13.30) are derived. Employing these results we find:

$$\begin{aligned} P^{(2)}\; v\; P^{(2)} &= (v_C+2v_T-v_{LS})\; P^{(2)} , \\ P^{(3)}\; v\; P^{(3)} &= (v_C-4v_T-2v_{LS})\; P^{(3)} , \\ P^{(2)}\; v\; P^{(3)} &= v_{LS}(\vec{L}\cdot\vec{S}+2)\; P^{(3)} , \\ P^{(3)}\; v\; P^{(2)} &= v_{LS}(\vec{L}\cdot\vec{S}+1)\; P^{(2)} , \end{aligned}$$

where v is given by (13.17). Returning to the evaluation of the matrix elements, we find:

$$\begin{aligned} M_V^{(\text{coupled})} = \sum_J 4\pi\int_0^\infty r^2 dr\; \{ &(v_C+2v_T-v_{LS})\left(f_J^{(2)}\right)^2\left((J+1)j_{J-1}^2 + J\, j_{J+1}^2\right) + \\ &+ (v_C-4v_T-2v_{LS})\left(f_J^{(3)}\right)^2\left(J\, j_{J-1}^2 + (J+1)j_{J+1}^2\right) + \\ &+ v_{LS}\; f_J^{(2)} f_J^{(3)}\, (2J)(J+1)\left(j_{J-1}^2 - j_{J+1}^2\right)\} . \end{aligned}$$

With the Reid potentials, v_C, v_T and v_{LS} are specified for each J wave individually.

The matrix elements for the T_1 term, denoted by:

$$M_T(ij) = \langle i,j|\; [f(12),\; [\nabla^2,\; f(12)]]\; |i,j\rangle \equiv M_T^{(\text{uncoupled})} + M_T^{(\text{coupled})} , \tag{13.32}$$

consist of again parts involving coupled and uncoupled states. The projection operators of the uncoupled states are not to be differentiated by the gradient operator, and therefore:

$$\begin{aligned} [\; f_J^{(0)} P_J^0,\; [\nabla^2,\; f_J^{(0)} P_J^0]\;] &= -2\left(\vec{\nabla} f_J^{(0)}\right)^2\; P_J^0 , \\ [\; f_F^{(1)} P_{J=L}^1,\; [\nabla^2,\; f_J^{(1)} P_{J=L}^1]\;] &= -2\left(\vec{\nabla} f_J^{(1)}\right)^2\; P_{J=L}^1 . \end{aligned} \tag{13.33}$$

Hence,

$$M_T^{(uncoupled)} = -\frac{2}{\Omega}\sum_J 4\pi(2J+1)\int_0^\infty r^2 dr\ j_J^2(kr)\{(f_J^{(0)\prime})^2 + (f_J^{(1)\prime})^2\}\ ,\quad (13.34)$$

where a prime denotes differentiation with respect to r. In analogy to (13.28), the matrix elements involving coupled states are given by:

$$M_T^{(coupled)} = \frac{1}{\Omega}\sum_J\sum_{L=J\pm1} 4\pi(2J+1)\int_0^\infty r^2 dr\ j_L^2(kr)\int d\Omega\ Y_{J1L}^{0}\ \times$$
$$\times[(f_J^{(2)}P^{(2)}+f_J^{(3)}P^{(3)}),[\nabla^2,(f_J^{(2)}P^{(2)}+f_J^{(3)}P^{(3)})]]\ Y_{J1L}^{0}.\quad (13.35)$$

The double commutator may be evaluated with straight forward manipulations:

$$[f_J^{(2)},[\nabla^2,f_J^{(2)}]]P^{(2)}+[f_J^{(3)},[\nabla^2,f_J^{(3)}]]P^{(3)}+(f_J^{(2)}-f_J^{(3)})^2[P^{(2)},[\nabla^2,P^{(2)}]]$$
$$= -2(\vec\nabla f_J^{(2)})^2P^{(2)} - 2(\vec\nabla f_J^{(3)})^2P^{(3)} - r^{-2}(f_J^{(2)}-f_J^{(3)})^2(4P_3-2P^{(2)}+2\vec L\cdot\vec S)\ ,$$

where the last line is evaluated with the assistance of the following results:

$$\nabla^2 P^{(2)} = r^{-2}(4P_3 - 6P^{(2)})\ ,$$
$$\vec\nabla P^{(2)}\cdot\vec\nabla = -r^{-2}(\vec\sigma_1\cdot\hat r)(\vec\sigma_2\cdot\hat r)\ \vec L\cdot\vec S\ ,\quad (13.36)$$

which are obtained in a manner similar to that leading to (13.33).

Thus,

$$M_T^{(coupled)} = \frac{1}{\Omega}\sum_J 4\pi\int r^2 dr\ \{-2(f_J^{(2)\prime})^2(J\ j_{J+1}^2 + (J+1)j_{J-1}^2) +$$
$$- 2(f_J^{(3)\prime})^2((J+1)j_{J+1}^2 + J\ j_{J-1}^2) - 4r^{-2}J(J+1)(f_J^{(2)}-f_J^{(3)})^2(j_{j-1}^2-j_{J+1}^2)\}.\quad (13.37)$$

Finally, the expectation value of the Hamiltonian in the lowest order approximation is evaluated to be:

$$\bar H = (0.6)\frac{\hbar^2}{2m}k_F^2 A + \frac{4\pi}{2\Omega}\sum_{i,j}\int r^2 dr\ \Big\{(2J+1)j_J^2(kr)\{\frac{\hbar^2}{m}(f_J^{(0)\prime})^2 + \frac{\hbar^2}{m}(f_J^{(1)\prime})^2 +$$
$$+ v_C(f_J^{(0)})^2 + (v_C+2v_T-v_{LS})(f_J^{(1)})^2\} +$$
$$+ (Jj_{J+1}^2(kr)+(J+1)j_{J-1}^2(kr))\{\frac{\hbar^2}{m}(f_J^{(2)\prime})^2 + (v_C+2v_T-v_{SL})(f_J^{(2)})^2\} +$$
$$+ ((J+1)j_{J+1}^2(kr)+Jj_{J-1}^2(kr))\{\frac{\hbar^2}{m}(f_J^{(3)\prime})^2 + (v_C-4v_T-2v_{LS})(f_J^{(3)})^2\} +$$
$$+ 2J(J+1)(j_{J-1}^2(kr)-j_{J+1}^2(kr))\{\frac{\hbar^2}{m}r^{-2}(f_J^{(2)}-f_J^{(3)})^2 + v_{LS}f_J^{(2)}f_J^{(3)}\}\Big\}\quad (13.38)$$

The summations over occupied states may be converted into integrals over momentum space as it is done in (12.46) and (12.50):

$$\sum_{i,j} \to \sum_{S,T} \Omega(2\pi)^{-3} \int 4\pi k_i^2 dk_i \gamma^{-1} n(k,k_i) dk$$
$$= \sum_{S,T} \Omega^2 (2\pi)^{-3} (8/\pi) \int k_i^2 dk_i \; k^2 dk \; \{1 + \eta(k,k_i)\} \; , \qquad (13.39)$$

where S,T stand for total spin and isospin, respectively, which are summed explicitly and hence the degeneracy factor γ is taken out of these expressions; $\eta(k,k_i)$ is given by (12.47). Since the matrix elements do not depend on k_i, integration over k_i may be carried out. Consider first $2k < k_F$, then:

$$\int k_i^2 dk_i \{1 + \eta(k,k_i)\} = \int_0^{(k_F-2k)} 2k_i^2 dk_i + \int_{(k_F-2k)}^{k_F} \{1 + \frac{k_F^2 - k_i^2 - 4k^2}{4kk_i}\} k_i^2 dk_i$$
$$= \frac{2}{3} k_F^3 \{1 - \frac{3}{2}\frac{k}{k_F} - \frac{1}{2}\left(\frac{k}{k_F}\right)^3\} \; , \qquad (13.40)$$

and for $2k > k_F$, the same result is obtained with:

$$\int k_i^2 dk_i \{1 + \eta(k,k_i)\} = \int_{(2k-k_F)}^{k_F} \{1 + \frac{k_F^2 - k_i^2 - 4k^2}{4kk_i}\} k_i^2 dk_i$$
$$= \frac{2}{3} k_F^3 \{1 - \frac{3}{2}\frac{k}{k_F} - \frac{1}{2}\left(\frac{k}{k_F}\right)^3\} \; . \qquad (13.41)$$

In short, the summations over occupied states are given by:

$$\sum_{i,j} \to \sum_{S,T} (A/\gamma)^2 \; 24 \int z^2 dz \; \{1 - \frac{3}{2} z + \frac{1}{2} z^3\} \; , \qquad (13.42)$$

where $z = k/k_F$. Inserting this expression into (13.38), the expectation value of the Hamiltonian in the lowest order approximation is given by:

$$\overline{H}/A = (0.6)\frac{\hbar^2}{2m} k_F^2 + \frac{2^{\pi} n_B}{\gamma^2 k_F^2} \sum_{J,L,S,T} (2T+1)(2J+1)\frac{1}{2}\{1 - (-1)^{L+S+T}\} \times$$
$$\times \int_0^{\infty} dr \left\{ \{\frac{\hbar^2}{m}\left(f_J^{(0)\prime}\right)^2 + v_C\left(f_J^{(0)}\right)^2 + \frac{\hbar^2}{m}\left(f_J^{(1)\prime}\right)^2 + (v_C + 2v_T - v_{LS})\left(f_J^{(1)}\right)^2\} a_J^{(1)2}(k_F r) \right. -$$
$$+ \{\frac{\hbar^2}{m}\left(f_J^{(2)\prime}\right)^2 + (v_C + 2v_T - v_{LS})\left(f_J^{(2)}\right)^2\} \; a_J^{(2)2}(k_F r) \; +$$

$$+ \left\{ \frac{\hbar^2}{m}\left(f_J^{(3)\prime}\right)^2 + (v_C - 4v_T - 2v_{LS})\left(f_J^{(3)}\right)^2 \right\} a_J^{(3)^2}(k_F r) \; +$$

$$+ \left\{ \frac{\hbar^2}{m} r^{-2}\left(f_J^{(2)} - f_J^{(3)}\right)^2 + v_{LS} f_J^{(2)} f_J^{(3)} \right\} b_J^2(k_F r) \Big\} \qquad (13.43)$$

where,

$$a_J^{(1)^2}(x) = x^2\, I_J(x) \; ,$$

$$a_J^{(2)^2}(x) = x^2 (2J+1)^{-1} \{ (J+1) I_{J-1}(x) + J\, I_{J+1}(x) \} \; ,$$

$$a_J^{(3)^2}(x) = x^2 (2J+1)^{-1} \{ J\, I_{J-1}(x) + (J+1) I_{J+1}(x) \} \; ,$$

$$b_J^2(x) = x^2\; 2J(J+1)(2J+1)^{-1} \{ I_{J-1}(x) - J_{J+1}(x) \} \; , \qquad (13.44)$$

with,

$$I_J(x) = 48 \int_0^1 dz\, z^2 \{ 1 - \tfrac{3}{2} z + \tfrac{1}{2} z^3 \}\, j_J^2(xz) \; . \qquad (13.45)$$

$\bar{H}/A$ given by (13.43) is also referred to as the two-body energy. In a review article on LOCV by Irvine (1981), (13.43) is the starting expression for the discussion, and we have followed this article's notations as much as possible. The two-body energy is now minimized with respect to variations in the correlation factors $f_J^{(i)}$, which as we have discussed before must be constrained so as to be consistent with the assumption of short-range correlations. There has been various different constraints employed. A reasonable set of constraints on the correlation factors are:

$$\text{(i)} \quad 0 \le f_J^{(i)}(r) \le 1 \quad \text{for} \quad r \le d_J^{(i)} \; ,$$

$$\text{and} \quad f_J^{(i)}(r) = 1 \quad \text{for} \quad r > d_J^{(i)} \; ; \qquad (13.46)$$

$$\text{(ii)} \quad \frac{4\pi n_B}{k_F^2} \sum_{J,L,S,T} \sum \int_0^\infty dr \sum_{i=1}^{3} \left(1 - f_J^{(i)^2}\right) a_J^{(i)^2}(k_F r) \le 1 \; . \qquad (13.47)$$

Constraint (i) prevents the long-range part of the potential to have a large effect on the correlation factors as it is essential to the success of the lowest order approximation. Constraint (ii) insures that the correlations do not expel more than one particle from within the correlation distance. It is also a consistency condition for a two-body approximation.

Variations of (H/A) with constraint (ii) give rise to the following set of equations for the correlation factors:

$$g_J^{(0)''} - \{(a_J^{(1)''}/a_J^{(1)}) + (m/\hbar^2)(v_C+\lambda)\}\, g_J^{(0)} = 0 ,$$

$$g_J^{(1)''} - \{(a_J^{(1)''}/a_J^{(1)}) + (m/\hbar^2)(v_C+2v_T-v_{LS}+\lambda)\}\, g_J^{(1)} = 0 ,$$

$$g_J^{(2)''} - \{(a_J^{(2)''}/a_J^{(2)}) + (m/\hbar^2)(v_C+2v_T-v_{LS}+\lambda) + (b_J/a_J^{(2)})^2/r^2\}\, g_J^{(2)} +$$

$$+ \{(r^{-2}-(m/2\hbar^2)v_{LS})(b_J^2/a_J^{(2)}a_J^{(3)})\}g_J^{(3)} = 0 ,$$

$$g_J^{(3)''} - \{(a_J^{(3)''}/a_J^{(3)}) + (m/\hbar^2)(v_C-4v_T-2v_{LS}+\lambda) + (b_J/a_J^{(3)})^2/r^2\}\, g_J^{(3)} +$$

$$+ \{(r^{-2}-(m/2\hbar^2)v_{LS})(b_J^2/a_J^{(2)}a_J^{(3)})\}g_J^{(2)} = 0 , \tag{13.48}$$

where λ denotes the Lagrange multiplier, and:

$$g_J^{(0)} = f_J^{(0)}(r)\, a_J^{(1)}(k_F r) ,$$

$$g_J^{(i)} = f_J^{(i)}(r)\, a_J^{(i)}(k_F r) , \quad \text{for } i = 1, 2, 3.$$

Constraint (i) will impose boundary conditions on $f_J^{(i)}$, or in this case on $g_J^{(i)}$. However, constraint (i) should be modified slightly for a Fermi system since antisymmetry in the two-body wave functions induces exclusion. In other words, the kinematical factors of the partial wave expansion do not summed up to unity but to a function of r, as follows:

$$(\gamma k_F r)^{-2} \sum_{J,L,S,T}\sum (2T+1)(2J+1)\tfrac{1}{2}\{1 - (-1)^{L+S+T}\}\sum_{i=1}^{3} a_J^{(i)2} = f_F(k_F r)^{-2}, \tag{13.49}$$

where,

$$f_F(x)^{-2} = \{ 1 - \frac{9}{\gamma}\left(\frac{j_1(x)}{x}\right)^2 \} \tag{13.50}$$

This result can be seen most easily from (7.39) by examining the part proportional to v_1. It gives:

$$\frac{1}{A}\sum_{i,j} \langle i,j \mid i,j - j,i\rangle = 2\pi n_B \int_0^\infty dr\, r^2\, f_F(k_F r)^{-2} , \tag{13.51}$$

which is the result of (13.49).

Thus, $f_J^{(i)}$ should be normalized to f_F instead of to unity as stated in (13.46). For $\gamma=4$, $f_F(k_F r) = 1.1547$ at $r=0$, and it drops to unity at $k_F r=4.37$, and oscillates with small amplitude about unity thereafter; f_F is therefore not too far different from unity. This particular normalization constraint has been adopted by Irvine and collaborators (see Irvine 1981).

To solve for $f_J^{(i)}$ or what is the same $g_J^{(i)}$, one integrates the equations of (13.48) out from the origin beginning with $f_J^{(i)}(0) = 0$ and arbitrary $f_J^{(i)\prime}(0)$, until the lograithmic derivatives of $f_J^{(i)}$ match those of f_F. The points of matching are the $d_J^{(i)}$ of (13.46). Trial solutions of $f_J^{(i)}$ should begin with $\lambda = 0$, and λ is increased if $f_J^{(i)}$ found do not satisfy (13.47). λ is then progressively incresed until (13.47) is satisfied. In conformity with discussion above, the quantity $(1 - f_J^{(i)2})$ in (13.47) should be modified to read $(f_F^2 - f_J^{(i)2})$. There are slight complications regarding the determination of d_J for $f_J^{(2)}$ and $f_J^{(3)}$ by this method, since they are coupled. We refer the interested readers to the review article by Irvine (1981) for a discussion of this point.

The status of the nuclear matter calculations employing LOCV method described here has been extensively reviewed in Irvine's article. We again refer the interested readers there for details. Briefly, LOCV calculations for nuclear matter with the Reid potentials as input yield a saturation energy of -23 MeV at a density given by $k_F=1.7\ fm^{-1}$. These are to be compared with the BBG results of Section 12. There the saturation energy from the Reid potentials is found to be about -12 MeV at $k_F=1.4\ fm^{-1}$. There are substantial differences in these results from these two methods. As we have mentioned before, the BBG method probably underestimates the binding energy by ignoring the three-body correlations. The LOCV method on the other hand may have made compensations for that by imposing physically realistic constraints on the correlation factors. However, since the constraint conditions are only chosen heuristically, it is difficult to pinpoint the source of uncertainly.

The LOCV method was first applied by Pandharipande (1971a, 1971b) to study neutron matter at densities near the nuclear density. The objective was to obtain the relevant equation of state for the neutron star. The interactions chosen for these works were obtained by neglecting the non-central components, such as v_T and v_{LS}, from the nucleon potentials, and using (i) the $^1D_2$ central potential for all even-L states except for the L=0 state which is treated with the $^1S_0$ potential, and (ii) the central part of the $(^3P_2-^3F_2)$ potential for all odd-L states. With these interactions, all partial waves are uncoupled. The variational equations for the partial wave correlation factors are:

$$-\frac{\hbar^2}{2m}\; f_J'' j_J + 2\, f_J' j_J' \;+ (v+\lambda) f_J\, j_J = 0 \;. \tag{13.51}$$

The boundary conditions are chosen to be:

$$f_J(0) = 0 \;,$$
$$f_J'(d) = 0 \;,$$
$$f_J(d) = 1 \;, \tag{13.52}$$

where d, called the healing distance, is fixed. These three boundary conditions turned the variational equations (13.51) into eigenvalue equations for λ, since the equations are over constrained. There will be one λ, designated by λ_J, for each wave. The choices of the healing distances are:

$$d^{\text{even } J} = 0.98\; r_0 \;,$$
$$d^{\text{odd } J} = 1.72\; r_0 \;, \tag{13.53}$$

where r_0 is the average nucleon separation, or $r_0=(3n_B/4\pi)^{-3}$. Results of this study employing several nuclear potentials: Reid, Bressel-Kerman-Rouben, and Hamada-Johnston, are discussed by Pandharipande (1971a). In a later version with somewhat modified boundary conditions, Pandharipande (1971b) reevaluated the equation of state for a pure neutron matter. This is shown in Figure 13-1. Owen, Bishop, Irvine (1976) choosing constraints similar to those discuss here obtained E/A values for neutron matter generally higher than those obtained by Pandharipande at densities above the nuclear densities. Possible reasons of discrepancy are discussed by Owen et al.

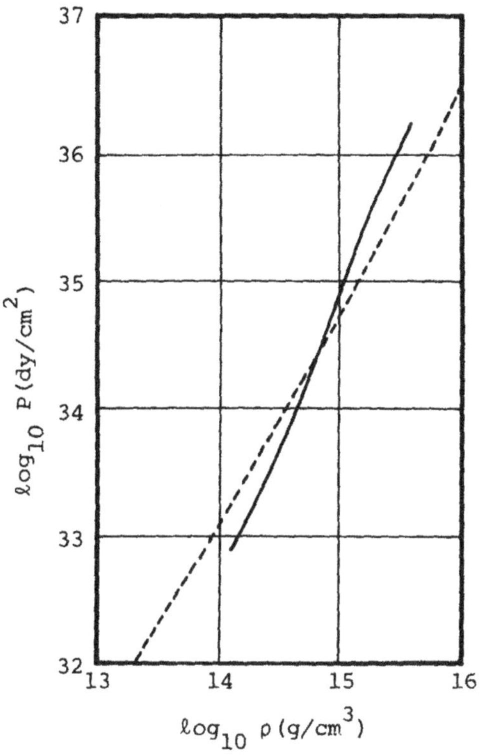

Figure 13-1. Equation of state for neutron matter obtained by Pandharipande (1971) by applying the variational method.

Recent study of nuclear and neutron matter by Friedman and Pandharipande (1980) introduces new interactions, which contain both two-nucleon and three-nucleon interactions. The equation of state obtained for neutron matter is found to be stiffer than that shown in Figure 13-1 This study includes finite temperature situations by computing the entropy and the free energy of the system. We shall not go into details here, but refer the interested readers to this article.

References

Clark, J.W. (1979). In Progress in Part. and Nucl. Phys., (Ed. D. Wilkinson). Pergamon Press, Oxford.

Friedman, B. and Pandharipande, V.R. (1981). Nucl. Phys. A361, 502.

Irvine, J. M. (1981). In Progress in Part. and Nucl. Phys. (Ed. D. Wilkinson), Vol. 5. Pergamon Press, Oxford.

Jastrow, R. (1955). Phys. Rev. 98, 1479.

Owen, J.C., Bishop, R.F. and Irvine, J.M. (1977). Nucl. Phys. A277, 45.

Pandharipande, V.R. (1971a). Nucl. Phys. A174, 641.

Pandharipande, V.R. (1971b). Nucl. Phys, A178,123.

Ristig, M.L., Ter Louw, W.J. and Clark,J.W. (1971). Phys. Rev. C3, 1504.

Ristig, M.L., Ter Louw, W.J., and Clark, J.W. (1972). Phys. Rev. C5, 695.

Bibliography

Guardiola, R. and Ros, J. (Eds.) (1981). The Many-Body Problem: Jastrow Correlations Versus Brueckner Theory, Springer-Verlag, Berlin and New York.

Zabolitzky, J.G., deLlano, M., Fortes, M. and Clark, J.W. (1981). Recent Progress in Many-Body Theories, Springer-Verlag, Berlin and New York.

14. Relativistic Models

Up till now we have been studying the nuclear problem entirely in the framework of non-relativistic theories. This is the traditional approach in nuclear physics and the justification for it is that the nuclear mass is large in comparison with either the kinetic energy or interaction energy involved. The maximum kinetic energy of a nucleon in nuclear matter is given by its Fermi energy in the range of 50 MeV, which is only one-twentieth of the nucleon mass. Similarly, the nuclear interactions as given by the Reid potentials are not excessively large at average nuclear separations. Hence, for nuclear problems non-relativistic theories can be justified. However, to study matter at transnuclear densities, relativistic theories become necessary. In preparation for this eventuality we shall examine here some relativistic models of nuclear matter. There are two aspects to the relativistic thoery: one is kinematical and the other dynamical. The kinematical aspect refers to the Lorentz covariance of its formulation, while the dynamical aspect refers to its explicit dynamical inputs, such as those expressed in meson theory or dispersion relations. In this section, we shall first look at the kinematical aspect of a relativistic theory of nucleons.

We shall adopt the following notational conventions. A Lorentz four-vector is written as:

$$b_\mu = (b_1, b_2, b_3, b_4) = (\vec{b}, ib_0) , \tag{14.1}$$

where b_0, b_1, b_2, and b_3 are real. In general we use Greek subscripts: μ, ν, λ, etc., running from 1 to 4 to denote 4-vector components, while Laten indices: i, j, k, etc., running from 1 to 3 to denote 3-vector components. When indices are repeated, summation over the full range of these indices are implied. The scalar product of four-vectors is defined by:

$$b \cdot c = b_\mu c_\mu = \vec{b}\cdot\vec{c} - b_0 c_0 . \tag{14.2}$$

For example, the space-time four-vector x_μ and the momentum-energy four vector p_μ are given by:

$$x_\mu = (\vec{x}, ict) ,$$
$$p_\mu = (\vec{p}, iE/c) ,$$

and their scalar products;

$$x\cdot x = x_\mu x_\mu = \vec{x}^2 - c^2t^2 ,$$
$$p\cdot p = p_\mu p_\mu = \vec{p}^2 - E^2/c^2 ,$$
$$p\cdot x = p_\mu x_\mu = \vec{p}\cdot\vec{x} - Et .$$

Also, we write;

$$\partial_\mu\partial_\mu = \vec{\nabla}^2 - \frac{1}{c^2}\frac{\partial^2}{\partial t^2} .$$

The free Dirac equation is written as;

$$i\hbar\frac{\partial\psi}{\partial t} = (-ic\hbar\,\vec{\alpha}\cdot\vec{\nabla} + \beta mc^2)\psi , \tag{14.3}$$

where $\vec{\alpha} = (\alpha_1, \alpha_2, \alpha_3)$ and β are 4×4 matrices satisfying the anticommutation relations:

$$\{\alpha_k, \alpha_\ell\} = 2\,\delta_{k\ell} \qquad \{\alpha_k, \beta\} = 0 , \quad \alpha_k^2 = \beta^2 = 1 , \tag{14.4}$$

where $\{A,B\} \equiv AB + BA$. Explicit realizations of $\vec{\alpha}$ and β are as follows:

$$\vec{\alpha} = \begin{pmatrix} 0 & \vec{\sigma} \\ \vec{\sigma} & 0 \end{pmatrix} , \qquad \beta = \begin{pmatrix} I & 0 \\ 0 & -I \end{pmatrix} , \tag{14.5}$$

where $\vec{\sigma}$ are the Pauli matrices and I is a 2x2 unit matrix. It is sometimes convenient to rewrite the Dirac equation in terms of the Dirac gamma matrices γ_μ defined to be:

$$\vec{\gamma} = -i\beta\vec{\alpha} , \qquad \gamma_4 = \beta . \tag{14.6}$$

They satisfy the anticommutation relations:

$$\{ \gamma_\mu, \gamma_\nu \} = 2\delta_{\mu\nu} , \tag{14.7}$$

and are hermitian , $\gamma_\mu^+ = \gamma_\mu$. In terms of the gamma matrices, the free Dirac equation is written as:

$$\{ \gamma_\mu\partial_\mu + (mc/\hbar) \} \psi = \{ \vec{\gamma}\cdot\vec{\nabla} + \gamma_4 \frac{\partial}{ic\partial t} + (mc/\hbar)\} \psi = 0 . \tag{14.8}$$

The solution ψ of (14.8) is expressed as a column matrix of four components;

$$\psi = \begin{pmatrix} \psi_1 \\ \psi_2 \\ \psi_3 \\ \psi_4 \end{pmatrix} \tag{14.9}$$

which is called a Dirac spinor. The Hermitian conjugate spinor ψ^+ is given by a row matrix;

$$\psi^+ = (\psi_1^*, \psi_2^*, \psi_3^*, \psi_4^*) . \tag{14.10}$$

For convenience we introduce also the adjoint spinor $\overline{\psi}$:

$$\overline{\psi} = \psi^+ \gamma_4 = (\psi_1^*, \psi_2^*, -\psi_3^*, -\psi_4^*) , \tag{14.11}$$

which satisfies the adjoint Dirac equation:

$$- \partial_\mu \overline{\psi} \gamma_\mu + (mc/\hbar) \overline{\psi} = 0 . \tag{14.12}$$

(14.12) is derived from (14.8) by making use of the relations:

$$\partial_4^* = - \partial_4 , \text{ and } \quad \gamma_k \gamma_4 = - \gamma_4 \gamma_k .$$

We shall henceforth adopt units in which $\hbar = c = 1$ with mass and momentum both expressed in energy units and length in inverse energy units. The plane wave solutions of the Dirac equation are expressed as:

$$\psi = \Omega^{-\frac{1}{2}} u(\vec{p}) \exp\{ i (\vec{p}\cdot\vec{x} - Et)\} . \tag{14.13}$$

Writing

$$u(\vec{p}) = \begin{pmatrix} u_A(\vec{p}) \\ u_B(\vec{p}) \end{pmatrix} ,$$

the free Dirac equation decomposes into:

$$\begin{aligned} \vec{\sigma}\cdot\vec{p}\, u_B &= (E - m)\, u_A , \\ \vec{\sigma}\cdot\vec{p}\, u_A &= (E + m)\, u_B , \end{aligned} \tag{14.14}$$

where $\vec{\sigma}\cdot\vec{p}$ may be written explicitly as:

$$\vec{\sigma}\cdot\vec{p} = \begin{pmatrix} p_z & p_- \\ p_+ & -p_z \end{pmatrix} , \tag{14.15}$$

with $p_\pm = p_x \pm ip_y$. As it is well-known, there are four independent solutions to the Dirac equation: two correspond to $E > 0$, and two others to $E < 0$. It is customary to write the solutions as follows: solutions 1 and 2, $E = p_0 \equiv (\vec{p}^2+m^2)^{\frac{1}{2}} > 0$:

$$u_A^{(1,2)}(\vec{p}) = N\,\chi_\sigma$$
$$u_B^{(1,2)}(\vec{p}) = \frac{\vec{\sigma}\cdot\vec{p}}{(p_0+m)}\,u_A^{(1,2)}(\vec{p}) \,, \tag{14.16}$$

where N is a normalization constant and $\chi_\sigma = \chi_\uparrow$ or $\chi_\downarrow$ are the spin wave functions of (7.2). Explicitly,

$$u^{(1)}(\vec{p}) = N\begin{pmatrix} 1 \\ 0 \\ p_z/(p_0+m) \\ p_+/(p_0+m)\end{pmatrix} \,, \qquad u^{(2)}(\vec{p}) = N\begin{pmatrix} 0 \\ 1 \\ p_-/(p_0+m) \\ -p_z/(p_0+m)\end{pmatrix} . \tag{14.17}$$

Solutions 3 and 4, $E = -p_0 < 0$:

$$u^{(3)}(\vec{p}) = N\begin{pmatrix} -p_z/(p_0+m) \\ -p_+/(p_0+m) \\ 1 \\ 0\end{pmatrix} \,, \qquad u^{(4)}(\vec{p}) = N\begin{pmatrix} -p_-/(p_0+m) \\ p_z/(p_0+m) \\ 0 \\ 1\end{pmatrix} . \tag{14.18}$$

The normalization condition is chosen to be:

$$u^{(r)+}\,u^{(r)} = 1 \,, \text{ (no summation over } r;\ r = 1, 2, 3, 4) \tag{14.19}$$

which means, $N = \sqrt{(p_0+m)/2p_0}$. (Note that it is sometimes desirable to express quantities in manifestly covariant forms, in which case a factor $\sqrt{m/p_0}$ is taken from $u(\vec{p})$ and combined with $\sqrt{1/\Omega}$ to form $\sqrt{m/p_0\Omega}$ which transforms like a Lorentz scalar; thus, the normalization of the spinors become $u^{(r)+}u^{(r)} = p_0/m$, as in Sakurai (1967), whereas our notations are similar to those of Rose (1961).) The wave function is therefore normalized to:

$$\int_\Omega \psi^+\,\psi\, d^3x = 1 \,. \tag{14.20}$$

The Dirac equation for a particle interacting with an external (time-independent) potential is written as:

$$\{-i\vec{\alpha}\cdot\vec{\nabla} + \beta(m + U)\}\psi = E\,\psi\ , \tag{14.21}$$

where the potential U will depend on the γ-matrices or products of γ-matrices. There are altogether 16 such independent Hermitian matrices and they are grouped together according to:

		Number of independent matrices
Γ_S :	I	1
Γ_V :	γ_μ	4
Γ_T :	$\sigma_{\mu\nu} = \frac{i}{2}(\gamma_\mu\gamma_\nu - \gamma_\nu\gamma_\mu)$	6
Γ_A :	$i\gamma_5\gamma_\mu$	4
Γ_P :	γ_5	1 (14.22)

where the subscripts of Γ have the following meaning: S = scalar, V = vector, T = tensor, A = axial vector and P = pseudoscalar. These designations refer to the transformation properties of these matrices in the form of bilinear densities, $\bar{\psi}\,\Gamma_a\,\psi = \psi^+\gamma_4\Gamma_a\psi$, under the Lorentz transformations. These matrices have the form:

$$I = \begin{pmatrix} I & 0 \\ 0 & I \end{pmatrix} ,$$

$$\gamma_k = \begin{pmatrix} 0 & -i\sigma_k \\ i\sigma_k & 0 \end{pmatrix} , \qquad \gamma_4 = \begin{pmatrix} I & 0 \\ 0 & -I \end{pmatrix} ,$$

$$\sigma_{ij} = \begin{pmatrix} \sigma_k & 0 \\ 0 & \sigma_k \end{pmatrix} (i,j,k \text{ cyclic}), \qquad \sigma_{k4} = \begin{pmatrix} 0 & \sigma_k \\ \sigma_k & 0 \end{pmatrix} ,$$

$$i\gamma_5\gamma_k = \begin{pmatrix} \sigma_k & 0 \\ 0 & -\sigma_k \end{pmatrix} . \qquad i\gamma_5\gamma_4 = \begin{pmatrix} 0 & iI \\ -iI & 0 \end{pmatrix} ,$$

$$\gamma_5 = \begin{pmatrix} 0 & -I \\ -I & 0 \end{pmatrix} . \tag{14.23}$$

The most general form of U is therefore:

$$U = U^S + U_\mu^V\gamma_\mu + U_{\mu\nu}^T\sigma_{\mu\nu} + U_\mu^A(i\gamma_5\gamma_\mu) + U^P\gamma_5\ , \tag{14.24}$$

where, and henceforth, the 4×4 identity matrix I will be suppressed. The potentials U^a would have Lorentz transformation properties in accordance with their superscripts. In evaluating the matrice elements $\psi^+\gamma_4\Gamma_a\psi$, those Γ_a which are non-diagonal would connect the spinor u_A to u_B. For positive energy spinors, u_B is smaller than u_A by a factor of $\vec{p}/(p_0+m)$, and since the nuclear problem is approximately non-relativistic, we would expect the Γ_a which are non-diagonal to play a relatively minor role to those which are diagonal.

In the many-body context, the many-body Hamiltonian will be generalized to be:

$$H = \sum_{i=1}^{A} \{-i\,\vec{\alpha}(i)\cdot\vec{\nabla}_i + \beta(i)m\} + \frac{1}{2}\sum_{i,j} v_{ij} , \tag{14.25}$$

where the Dirac matrices indexed by i are to act on the spinor for the ith particle with coordinates $\vec{x}_i$, and the interaction potentials v_{ij} will be time-independent. Thus, a common time is chosen for all particles in the system and the description for all particles is to be restricted to this specific Lorentz frame.

The Hartree-Fock wave function for the system will be written as (1.17). It is not difficult to see that the Hartree-Fock equations for a system of identical Fermions are given by:

$$\{-i\vec{\alpha}(1)\cdot\vec{\nabla}_1 + \beta(1)m\}\psi_i(\vec{x}_1) + \sum_{j=1}^{A}\int d^3x_2\psi_j^+(\vec{x}_2)v(\vec{x}_1-\vec{x}_2)\{\psi_j(\vec{x}_2)\psi_i(\vec{x}_1) + \\ -\delta(\sigma_i,\sigma_j)\psi_i(\vec{x}_2)\psi_j(\vec{x}_1)\} = \varepsilon_i^{HF}\psi_i(\vec{x}_1) . \tag{14.26}$$

Here the single-particle Hartree-Fock energies ε^{HF} will be relativistic energies containing the mass of the particle.

For a homogeneous system, the spatial part of the single-particle wave function takes on the form of a plane wave as in (14.13). The Hartree-Fock equations reduce to:

$$\{\vec{\alpha}\cdot\vec{p} + \beta m\}\, u(\vec{p}) + \sum_{j=1}^{A}\Omega^{-1}\int d^3x_2\; u^+(\vec{q})\; v(\vec{x}_1-\vec{x}_2)\; u(\vec{q})\; u(\vec{p}) + \\ - \sum_{j=1}^{A}\delta(\sigma_i,\sigma_j)\Omega^{-1}\int d^3x_2\; e^{i(\vec{q}-\vec{p})\cdot(\vec{x}_1-\vec{x}_2)}\, u^+(\vec{q})v(\vec{x}_1-\vec{x}_2)u(\vec{p})u(\vec{q}) \\ = \varepsilon_p^{HF}\; u(\vec{p}) , \tag{14.27}$$

where we have removed the indices from the Dirac matrices with the understanding that they act on spinors next to them; the momentum of state i is designated by $\vec{p}$, while those of j states are designated by $\vec{q}$. The summation over j states may be converted into an integration over the momentum space:

$$\sum_{j=1}^{A} \rightarrow \sum_{\substack{\text{spin} \\ \text{isospin}}} (2\pi)^{-3} \int d^3q \ , \tag{14.28}$$

An interesting model of nuclear matter introduces the following form of relativistic effective interactions (Durrer 1956, Walecka 1974):

$$v(\vec{x}_1-\vec{x}_2) = -\beta(1)\beta(2)v_\sigma + \{I(1)I(2) - \vec{\alpha}(1)\cdot\vec{\alpha}(2)\}\ v_\omega \ , \tag{14.29}$$

where v_σ and v_ω depend on the relative coordinates and momenta of these particles. In terms of the classification given in (14.22), these interaction terms are proportional to $(\gamma_4\Gamma_S)(\gamma_4\Gamma_S)$ and $(\gamma_4\Gamma_V)(\gamma_4\Gamma_V)$. The first term designated by the subscript σ is to be associated with the interaction due to the exchange of a neutral scalar meson, called the σ-meson, in relativistic meson theory, while the second term designated by subscript ω arises out of the exchange of a neutral vector meson called the ω-meson. With such interactions, the Hartree-Fock equations are written as:

$$\{\vec{\alpha}\cdot\vec{p} + \beta(m + U^H + U^F)\}\ u(p) = \varepsilon_p^{HF}\ u(p) \ , \tag{14.30}$$

where U^H represents the direction interaction (or Hartree) term, and U^F the exchange interaction (or Fock) term, with:

$$\beta U^H = (\gamma/2)(2\pi)^{-3}\int d^3q \int d^3x_2 \sum_{r=1,2} \{ \left(u^{(r)+}(\vec{q})\ \beta u^{(r)}(\vec{q})\right) v_\sigma \beta + \\ + \left(u^{(r)+}(\vec{q})\ u^{(r)}(\vec{q})\right) - \left(u^{(r)+}(\vec{q})\ \vec{\alpha} u^{(r)}(\vec{q})\right) v_\omega \vec{\alpha}\} \tag{14.31}$$

and,

$$\beta U^F = -(2\pi)^{-3}\int d^3q \int d^3x_2\ e^{i(\vec{q}-\vec{p})\cdot(\vec{x}_1-\vec{x}_2)} \Big\{ -\left(\beta u^{(r)}(\vec{q})\right)\left(u^{(r)+}(\vec{q})\beta\right) v_\sigma + \\ + \{\left(u^{(r)}(\vec{q}) u^{(r)+}(\vec{q})\right) - \left(\vec{\alpha} u^{(r)}(\vec{q})\right)\left(u^{(r)+}(\vec{q})\vec{\alpha}\right)\} v_\omega \Big\} . \tag{14.32}$$

Let us first consider the Hartree equations by ignoring U^F. For a uniform system U^H can be decomposed as follows:

$$\beta U^H = \beta U_S^H + U_0^H . \tag{14.33}$$

The fact that it should contain no term proportional to $\vec{\alpha}$ is clear, since for an isotropic system no vectorial quantity of that type should emerge. In any case, (14.33) may be verified by direct evaluation after the spinors are known. The Hartree equations are therefore given by:

$$\{\vec{\alpha}\cdot\vec{p} + \beta(m + U_S^H) + U_0^H\}u(\vec{p}) = \varepsilon_p^H \, u(\vec{p}) . \tag{14.34}$$

In analogy to (14.16) the spinor solutions are given by:

$$u^{(1)}(\vec{p}) = N_H \begin{pmatrix} 1 \\ 0 \\ p_z/(p_H+m_H) \\ p_+/(p_H+m_H) \end{pmatrix} , \tag{14.35}$$

where,

$$m_H = m + U_S^H ,$$

$$p_H = (\vec{p}^2+m_H^2)^{\frac{1}{2}}$$

$$N_H = \sqrt{(p_H+m_H)/2p_H} . \tag{14.36}$$

A similar expression is obtained by $u^{(2)}(\vec{p})$. U^H of (14.31) is to be evaluated with these spinor solutions. For example,

$$\sum_{r=1,2} u^{(r)+}(\vec{q}) \, \beta \, u^{(r)}(\vec{q}) = (q_H+m_H)q_H^{-1}\{1 - \vec{q}^2(q_H+m_H)^{-2}\} = 2m_H/q_H ,$$

$$\sum_{r=1,2} u^{(r)+}(\vec{q}) \, u^{(r)}(\vec{q}) = (q_H+m_H)q_H^{-1}\{1 + \vec{q}^2(q_H+m_H)^{-2}\} = 2 ,$$

$$\sum_{r=1,2} u^{(r)+}(\vec{q}) \, \vec{\alpha} \, u^{(r)}(\vec{q}) = 2\vec{q}/q_H , \tag{14.37}$$

The last term in (14.37) being proportional to $\vec{q}$ vanishes upon symmetric integration over $\vec{q}$ and will not contribute to U^H. We then find:

$$U_S^H = - n_s \int d^3x_2 \, v_\sigma(|\vec{x}_1-\vec{x}_2|) , \tag{14.38}$$

where,

$$n_s = (\gamma/2\pi^2)\int_0^{k_F} q^2 dq \frac{m_H}{\sqrt{\vec{q}^2+m_H^2}}$$

$$= \gamma(16\pi^2 m_H^3)^{-1}\left\{(1+x^2)^{3/2} + (1+x^2)^{1/2} - \frac{1}{2}x^4 \ln\frac{(1+x^2)^{1/2}+1}{(1+x^2)^{1/2}-1}\right\} \quad (14.39)$$

with $x = k_F/m_H$, and:

$$U_0^H = n_B \int d^3x_2 \, v_\omega(|\vec{x}_1-\vec{x}_2|) \ , \quad (14.40)$$

where n_B is the nucleon number density.

If v_σ and v_ω are given by the Yukawa potentials:

$$v_a(r) = (g_a^2/4\pi)\frac{e^{-m_a r}}{r} \ , \quad (a=\sigma,\omega) \quad (14.41)$$

then,

$$\int d^3x_2 \, v_a(|\vec{x}_1-\vec{x}_2|) = (g_a^2/m_a^2) \ . \quad (14.42)$$

We deduce finally,

$$U_S^H = -n_s(g_\sigma^2/m_\sigma^2) \ , \quad (14.43)$$

$$U_0^H = n_B(g_\omega^2/m_\omega^2) \ . \quad (14.44)$$

We note that n_s is a function of m_H, which as given by (14.36) has the meaning of an effective mass and is again related back to n_s via U_S^H. Thus, at each matter density (or a fixed k_F), m_H is determined entirely by the σ-meson parameters, its coupling constant g_σ and mass m_σ, through the self-consistent relation:

$$m_H - m = \frac{g_\sigma^2}{m_\sigma^2}\frac{\gamma}{2\pi^2}\int_0^{k_F} q^2 dq \frac{m_H}{\sqrt{\vec{q}^2+m_H^2}} \ . \quad (14.45)$$

The spectrum of Hartree energies ε_p^H is obtained from (14.34) to be:

$$\varepsilon_p^H = U_0^H + \sqrt{\vec{p}^2 + m_H^2} \ . \quad (14.46)$$

The total energy E or the energy density E/Ω of the system is given by (1.16) to be:

$$E/\Omega = (\gamma/2)(2\pi)^{-3}\int d^3p \, \{2\varepsilon_p^H - \frac{1}{2}\sum_{r=1,2} u^{(r)+}(\vec{p})\left(U_0^H + \beta U_S^H \right)u^{(r)}(\vec{p})\}$$

$$= \frac{1}{2}(g_\sigma^2/m_\sigma^2)n_s^2 + \frac{1}{2}(g_\omega^2/m_\omega^2)n_B^2 + (\gamma/2\pi^2)\int_0^{k_F} p^2 dp\sqrt{\vec{p}^2+m_H^2} \, . \tag{14.47}$$

The binding energy per particle B is given by:

$$B = m - (E/A) \, , \tag{14.48}$$

where $A = n_B\Omega$. B is completely determined once the meson parameters are known. In this model, the meson parameters required are two: (g_σ^2/m_σ^2) and (g_ω^2/m_ω^2). Since the interaction potentials are introduced as effective interactions, we might use the empirical saturation properties of nuclear matter to determine the meson parameters. Imposing the conditions:

(i)
$$\frac{d(E/A)}{dk_F} = 0 \qquad \text{at } k_F = 1.42 \text{ fm}^{-1}, \tag{14.49}$$

and (ii) B = 15.75 MeV at that density, Walecka(1974) found that:

$$g_\sigma^2(m/m_\sigma)^2 = 267,$$
$$g_\omega^2(m/m_\omega)^2 = 196, \tag{14.50}$$

where m denotes the nucleon mass. With these parameters the model is completely specified, which is often referred to as the Walecka model. At nuclear density, the effective mass is found to be $m_H = 0.56$ m, or interaction energies (in units of nucleon mass) of $U_S^H = 0.44$ m and $U_0^H = 0.59$m. These results show that the interaction energies are comparable to the nucleon mass, and such large interaction energies would make non-relativistic treatment of the nuclear problem difficult to justify.

Since both the σ-meson and ω-meson are assumed to have zero isospin, they couple equally to protons and neutrons. The same expressions apply to neutron matter if the degeneracy factor is taken to be $\gamma=2$ instead of $\gamma=4$ for symmetric nuclear matter. Employing meson parameters given by (14.50), the per nucleon energies for both nuclear matter and neutron matter can be evaluated and are shown in Figure 14-1.

The pressure generated by the system is found from (4.24) to be:

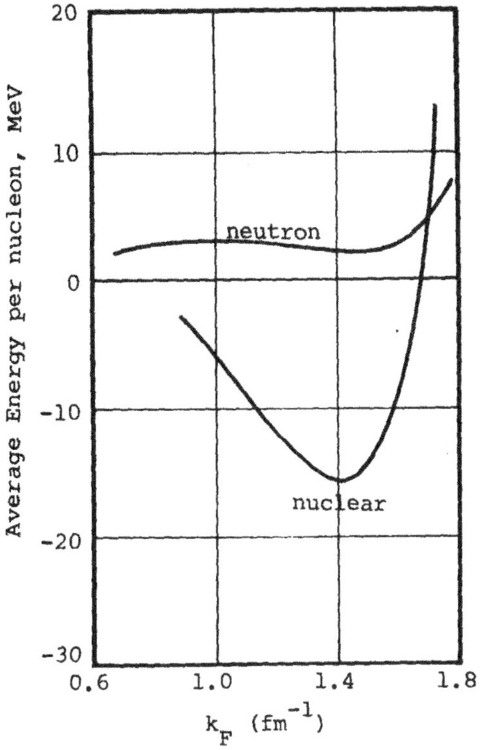

Figure 14-1. Average energies per nucleon for nuclear and neutron matter from the Walecka model.

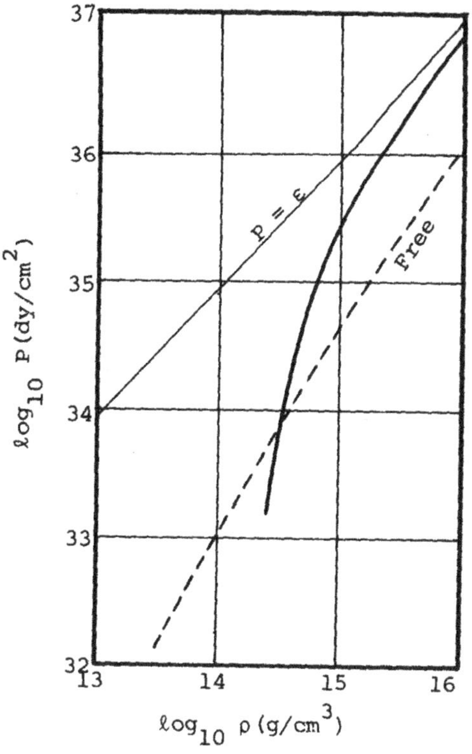

Figure 14-2. The equation of state for neutron matter computed from the Walecka model. (solid line).

$$P = -\frac{1}{2}(g_\sigma^2/m^2)n_s^2 + \frac{1}{2}(g_\omega^2/m^2)n_B^2 + (\gamma/6\pi^2)\int_0^{k_F}\frac{p^4 dp}{(\vec{p}^2+m_H^2)^{1/2}} \qquad (14.51)$$

The equation of state of neutron matter computed from this model is shown in Figure 14-2. Added for comparison there are the equation of state of free neutrons and that given by $P = \varepsilon$ $(=E/\Omega)$, for which the speed of sound equals the speed of light. The pressure given by the equation of state computed from the Walecka model in the density range 10^{14} - 10^{15} g/cm^3 is generally higher than that given by those discussed in the last section, but not unreasonable in view of the fact that the model employs only two phenomenological parameters. An equation of state which extends above the line $P = \varepsilon$ indicated in Figure 14-2 would have superluminal sound speed, which is impossible. Hence the $P = \varepsilon$ line is called the causality limit, which no equation of state should exceed. This criterium is always satisfied by relativistic models of particle interaction. When the equation of state from this model is extended to very high densities, it tends asymptotically to the $P = \varepsilon$ line. This is a result first obtained by Zel'dovich (1961).

This model is known to be imperfect. It predicts a nuclear compression modulus K (see 8.14) which is much too high. Empirically, the compression modulus for nuclear matter is estimated to be $K = 220 \pm 30$ MeV (Blaizot 1980). This model predicts $K = 550$ MeV.

In order to improve on the model, one might consider the inclusion of other mesons into the model. The most prominent candidate in this regard is the vector-isovector ρ-meson, which has a mass very close to that of the ω-meson. This was considered by Serot (1979). The interaction generated by the exchange of the ρ-meson however does not contribute to the nuclear matter energy which has zero net isospin, and therefore would not alter any of the results obtained without it. It will affect the neutron matter results.

In this model, the σ-meson generates a long-range attractive interaction, while the ω-meson generates a short-range repulsive interaction. Together, they accout for nuclear saturation, but unlike the ω-meson which is known to be a well-defined three-pion resonance, no σ-meson has actually been identified as such, and thus the form of v_σ given in (14.40) may not be realistic. There are proposed modifications of the term v_σ by introducing non-linear effects (Boguta and Bodmer 1977). We shall postpone discussion of these modifications until the next section, when the field theory method is introduced. It is found that by incorporating non-linear terms with additional adjustable parameters, it is possible to obtain a compression modulus in agreement with empirical value. Unfortunately, the correctness of this approach cannot be independently verified.

We shall next examine the complete Hartree-Fock equations. U^F given by (14.32) will in general be dependent on the momentum $\vec{p}$. It may be expressed as:

$$\beta U^F(\vec{p}) = \beta U_S^F(p) + U_0^F(p) + \vec{\alpha}\cdot\hat{p}\, U_V^F(p) , \qquad (14.52)$$

where U_S^F, U_0^F and U_V^F are scalar functions of p. The form given by (14.52) can be verified directly after the Hartree-Fock single-nucleon wave functions are found. The Hartree-Fock equations (14.30) are now written as:

$$\{\vec{\alpha}\cdot\vec{p}_v + \beta m_{HF} + U_0^{HF}\}\, u(\vec{p}) = \varepsilon_p^{HF}\, u(\vec{p}) \, , \tag{14.53}$$

where,

$$\vec{p}_v = \vec{p}\,(1 + \frac{1}{p} U_V^F) \, ,$$

$$m_{HF} = m + U_S^H + U_S^F \, ,$$

$$U_0^{HF} = U_0^H + U_0^F \, . \tag{14.54}$$

The spinor solutions are given by:

$$u^{(1)}(\vec{p}) = N_{HF} \begin{pmatrix} 1 \\ 0 \\ p_{vz}(p_{HF}+m_{HF})^{-1} \\ p_{v+}(p_{HF}+m_{HF})^{-1} \end{pmatrix} \, , \tag{14.55}$$

where,

$$p_{HF} = (\vec{p}_v^{\,2} + m_{HF}^2)^{1/2} \, ,$$

$$N_{HF} = \sqrt{(p_{HF}+m_{HF})/2p_{HF}} \, . \tag{14.56}$$

Referring to βU^F given by (14.32), we note that quantities like $\{\beta u^{(r)}\}\{u^{(r)+}\beta\}$ are given by 4×4 matrices. They can be evaluated as follows. Let us rewrite the spinors as:

$$u^{(r)}(\vec{q}) = N_{HF}(q_{HF}+m_{HF})^{-1} \begin{pmatrix} \chi_\sigma \\ \vec{\sigma}\cdot\tilde{\vec{q}}_V \chi_\sigma \end{pmatrix} \, , \tag{14.57}$$

then,

$$\{\beta u^{(r)}(\vec{q})\}\{u^{(r)+}(\vec{q})\beta\} = N_{HF}^2 (q_{HF}+m_{HF})^{-2} \begin{pmatrix} (q_{HF}+m_{HF})\chi_\sigma \\ -\vec{\sigma}\cdot\vec{q}_v\, \chi_\sigma \end{pmatrix} \begin{pmatrix} \chi_\sigma^+ (q_{HF}+m_{HF}) \, , & -\chi_\sigma^+ \vec{\sigma}\cdot\vec{q}_v \end{pmatrix}$$

$$= (2q_{HF})^{-1} \begin{pmatrix} (q_{HF}+m_{HF}) & -\vec{\sigma}\cdot\vec{q}_v \\ -\vec{\sigma}\cdot\vec{q}_v & (q_{HF}-m_{HF}) \end{pmatrix} \chi_\sigma \chi_\sigma^+$$

$$= (2q_{HF})^{-1} \{ I q_{HF} + \beta m_{HF} - \vec{\alpha}\cdot\vec{q}_v \} \chi_\sigma \chi_\sigma^+ \, , \tag{14.58}$$

where we have made use of the relation, $q_v^2 = (q_{HF}+m_{HF})(q_{HF}-m_{HF})$.

Similarly,

$$u^{(r)}(\vec{q})\, u^{(r)+}(\vec{q}) = (2q_{HF})^{-1} \{ I q_{HF} + \beta m_{HF} + \vec{\alpha}\cdot\vec{q}_v \} \chi_\sigma \chi_\sigma^+ \tag{14.59}$$

and,

$$\left(\vec{\alpha}u^{(r)}(\vec{q})\right)\left(u^{(r)+}(\vec{q})\vec{\alpha}\right) = (2q_{HF})^{-1}\{3Iq_{HF} - 3\beta m_{HF} - \vec{\alpha}\cdot\vec{q}_V\}\chi_\sigma\chi_\sigma^+ \quad (14.60)$$

where $\chi_\sigma\chi_\sigma^+$ acts as a projection operator for the spin-σ state and may be ignored since $u^{(r)}(\vec{p})$ on which it acts has the same spin component; this also allows $\chi_\sigma\chi_\sigma^+$ to commute by $\vec{\sigma}\cdot\vec{q}_V$. Thus,

$$U_S^F = (2\pi)^{-3}\int d^3q \int d^3x_2\; e^{i(\vec{p}-\vec{q})\cdot(\vec{x}_2-\vec{x}_1)} \frac{m_{HF}}{2\sqrt{\vec{q}_V^{\,2}+m_{HF}^2}}\{\, v_\sigma - 4v_\omega \,\},$$

$$U_0^F = (2\pi)^{-3}\int d^3q \int d^3x_2\; e^{i(\vec{p}-\vec{q})\cdot(\vec{x}_2-\vec{x}_1)} \frac{1}{2}\{\, v_\sigma + 2v_\omega \,\},$$

$$U_V^F = (2\pi)^{-3}\int d^3q \int d^3x_2\; e^{i(\vec{p}-\vec{q})\cdot(\vec{x}_2-\vec{x}_1)} \frac{\hat{p}\cdot\vec{q}_V}{2\sqrt{\vec{q}_V^{\,2}+m_{HF}^2}}\{\, -v_\sigma - 2v_\omega \,\}. \quad (14.61)$$

The exchange potential from meson theory is given by:

$$v_a(r) = \frac{g_a^2}{4\pi}\,\frac{1}{r}\,\exp\{-\sqrt{m_a^2-(p_0-p_0')^2}\, r\}\,, \quad (14.62)$$

where p_0 and p_0' are, respectively, the initial and final energies of the nucleon (Brockmann 1978). (14.62) reduces to the Yukawa form (14.41) for direct interaction when the initial and final energies of the nucleon are the same. For exchange interaction the initial energy is p_0 while the final energy is q_0. Inserting this form of the potential into (14.32) and employing spinor (14.57), we find:

$$U_S^F(p) = \frac{1}{2(2\pi)^3}\int d^3q\, \frac{m_{HF}}{\sqrt{\vec{q}_V^{\,2}+m_{HF}^2}}\{\, G(\sigma) - 4G(\omega)\,\}\,,$$

$$U_0^F(p) = \frac{1}{2(2\pi)^3}\int d^3q\, \{\, G(\sigma) + 2G(\omega)\,\}\,,$$

$$U_V^F(p) = \frac{1}{2(2\pi)^3}\int d^3q\, \frac{\hat{p}\cdot\vec{q}_V}{\sqrt{\vec{q}_V^{\,2}+m_{HF}^2}}\{-G(\sigma) - 2G(\omega)\,\}\,, \quad (14.63)$$

where,

$$G(a) = g_a^2\,\{m_a^2 + (p-q)_\mu (p-q)_\mu\}^{-1}\,, \qquad a=(\sigma,\omega)\,. \quad (14.64)$$

The Hartree-Fock effective mass m_{HF} depends on the magnitude of the momentum p. For every fixed values of p and k_F, it can be solved from the following self-consistency condition:

$$m_{HF} - m = \frac{g_\sigma^2}{m_\sigma^2}\frac{\gamma}{2\pi^2}\int_0^{k_F} q^2 dq \frac{m_{HF}}{\sqrt{\vec{q}_v^{\,2}+m_{HF}^2}} + \frac{\gamma}{2(2\pi)^3}\int d^3q \frac{m_{HF}}{\sqrt{\vec{q}_v^{\,2}+m_{HF}^2}}\{G(\sigma)-4G(\omega)\}. \tag{14.65}$$

The Hartree-Fock single-nucleon energy spectrum is given by:

$$\varepsilon_p^{HF} = U_0^{HF} + \sqrt{\vec{p}_v^{\,2}+m_{HF}^2}\ . \tag{14.66}$$

By including the exchange terms, the model depends on four meson parameters, namely, g_σ^2, m_σ^2, g_ω^2, and m_ω^2. Among these m_ω is most accurately known, $m_\omega = 782.8$ MeV, and conventionally, m_σ is taken to be $m_\sigma = 550$ MeV. With these meson masses (14.50) gives $g_\sigma^2/4\pi = 6.57$ and $g_\omega^2/4\pi = 9.25$. Jamion, Mahaux and Rochus (1981) have investigated this model for nuclear matter. They find that the exchange interactions have significant contribution to the average binding of the nucleons, which is substantially reduced compared to the Walecka model. In order to recover the empirical saturation properties the meson coupling constants must be increased by about 10% while keeping the meson masses to be the same. The meson parameters for the Hartree-Fock model are found to be:

$$m_\sigma = 550 \text{ MeV}, \qquad g_\sigma^2/4\pi = 7.47,$$

$$m_\omega = 782.8 \text{ MeV}, \qquad g_\omega^2/4\pi = 10.15. \tag{14.67}$$

Jamion et al. have further included cut-off parameters in their potentials which we have ignored here for clarity in presentation. These parameters should not alter the conclusions in any way.

References

Blaizot, J.P. (1980). Phys. Rep. 64, 171.

Boguta, J. and Bodmer, A.R. (1971). Nucl. Phys. A292, 413.

Brockmann, R. (1978). Phys. Rev. C18, 1510.

Durrer, Hans-Peter (1956). Phys. Rev. 103, 469.

Jaminon, M., Mahaux, C. and Rochus, P. (1981). Nucl. Phys. A365, 371.

Rose, M.E. (1961). Relativistic Electron Theory, Wiley, New York.

Sakurai, J.J. (1967). Advanced Quantum Mechanics, Addison-Wesley, Reading, Mass.

Serot, B.D. (1979). Phys. Lett. 86B, 146.

Walecka, J.D. (1974). Ann. Phys. 83, 491.

Zel'dovich, Ya. B. (1961). Jl. Exptl. Theoret. Phys. (U.S.S.R.) 41, 1609. (English transl., Sov. Phys. JETP 14, 1143, 1962).

Bibliography

Cottingham, W.N. (1978). Rep. Prog. Phys. 41, 1509.

Rho, M. and Wilkinson, D. (Eds.) (1979). Mesons in Nuclei, Vol. 1, North Holland, Amsterdam.

CHAPTER 4

Regime IV: $\rho > 10^{16}$ g/cm^3. Ultradense Matter.

15. Meson Theories

The structure of matter at densities much higher than nuclear density is largely a matter of theoretical speculations. There is hardly any direct empirical results available, although future successful interpretation of the experimental heavy ion collision data may provide the much needed information. Since the study of neutron stars requires the knowledge of the equation of state of matter at such high densities, much theoretical effort has been directed towards an understanding of the structure of matter at these densities. We summarize in this chapter some past efforts in this direction. We shall concentrate mainly on the following topics: (1) baryonic matter, (2) pion condensation, and (3) quark matter. The nature of these topics will be explained more fully in the following sections. Since much of the discussion from now on depends on the explicit nature of nucleon interactions, it is necessary first to recapitulate the results of the meson theory. The notations and formulations introduced will also be useful for the study of the quark matter.

In the previous sections, the particles are assumed to interact via potentials. This form of discussion has its limitations, since it does not explain the origin of the potentials. An attempt to understand the nature of nucleon interactions is the formulation of the meson theory. It postulates that particle interactions arise out of the exchange of field quanta called mesons. The meson theory is formulated in terms of an operator formalism instead of the wave function formalism that we have employed up to this point. We shall call the operator formalism the field theory, which employs a method known as the second quantization. Its importance derives from the fact that it keeps track of particle statistics at all stages of a calculation,and it allows formulation in which the number of particles is not a constant of the motion. Such a formulation is especially suitable for perturbative methods. However, since perturbative methods are not particularly successful in strong interaction physics, we have been resorting to effective interactions and phenomenological approaches, in which case the wave function approach has its appeals.

In the wave function formulation, the wave functions form the basis of a Hilbert space. It is called the Schrodinger representation. In field theory, the Hilbert space is spanned by a complete set of abstract vectors, or the occupation number space. The abstract vectors are denoted by $|\lambda>$ We do not need to know the explicit realizations of these vectors in terms of mathematical functions. They are presumably the eigenstates of a set of commuting operators. The orthonormality of the states is expressed by the notation:

$$<\lambda|\lambda'> = \delta_{\lambda\lambda'} \quad , \tag{15.1}$$

where $\delta_{\lambda\lambda'}$ denotes the Kronecker delta. Any operator $\mathcal{O}$ on such a space may be expressed as:

$$\mathcal{O} = \sum_{\lambda} c_{\lambda} \; |\lambda><\lambda| \quad , \tag{15.2}$$

and the completeness of the states is expressed by:

$$\sum_{\lambda} |\lambda><\lambda| = 1 \; . \tag{15.3}$$

It is convenient to think of $|\lambda>$ to be the eigenstates of a single-particle Hamiltonian with eigenvalues denoted by λ. A second Hilbert space of a similar type may consist of the eigenstates of the translational operator $\vec{T}$ (see 2.2). Such states are usually denoted by $|\vec{x}>$, and since there are nondenumerable number of such states the completeness condition is written as:

$$\int d^3x \;\; |\vec{x}><\vec{x}| = 1 \; . \tag{15.4}$$

Single-particle wave functions in the Schrodinger representation are given in the present notation by:

$$\psi_{\lambda}(\vec{x}) = <\vec{x}|\lambda> \tag{15.5}$$

Thus, (15.1) corresponds to the following statement in the Schrodinger representation:

$$\langle\lambda|\lambda'\rangle = \int d^3x \langle\lambda|\vec{x}\rangle\langle\vec{x}|\lambda'\rangle = \int d^3x\, \psi^+_\lambda(\vec{x})\psi_{\lambda'}(\vec{x}) = \delta_{\lambda\lambda'} \qquad (15.6)$$

A many-particle system may be obtained by stating the probability amplitudes for finding n_1 particles in the eigenstate $|\lambda_1\rangle$, n_2 in $|\lambda_2\rangle$, etc., with $\Sigma n_i = A$, which is the total number of particles in the systen We shall concentrate on the problem of a system of identical particles for now, since non-identical particles no not transform into each other and they occupy separate Hilbert spaces in the occupation number space formulation. Thus, for a many-particle state, we write:

$$|\Psi(t)\rangle = \sum_{n_1,n_2,\ldots n_\infty} f(n_1,n_2,\ldots,n_\infty,t)|n_1n_2\ldots n_\infty\rangle\ , \qquad (15.7)$$

where,

$$|n_1n_2\ldots n_\infty\rangle = |n_1\rangle|n_2\rangle \ldots\ldots |n_\infty\rangle\ ,$$

and

$$|n_i\rangle = |\lambda_i\rangle|\lambda_i\rangle \ldots\ldots\ldots |\lambda_i\rangle \qquad (n_i \text{ times})\ , \qquad (15.8)$$

is a direct product of A single-particle $|\lambda\rangle$ states. n_i is the occupation number in the state i. Note that each set of $|\lambda\rangle$ states spans a Hilbert space H and the set $|n_1n_2\ldots n_\infty\rangle$ states spans a Hilbert space $H^{(A)} = H \oplus H \oplus H \oplus \ldots\ldots$ (A times).

Next, we introduce the annihilation operator a_i and creation operator a_i^+ with the property that:

$$a_i\,|\,n_1n_2\ldots n_i\ldots\ldots\rangle = c_1|\,n_1n_2\ldots n_i-1\ldots\ldots\rangle\ , \qquad (15.9)$$

$$a_i^+\,|\,n_1n_2\ldots n_i\ldots\ldots\rangle = c_2|\,n_1n_2\ldots n_i+1\ldots\ldots\rangle\ , \qquad (15.10)$$

where c_1 and c_2 are real numbers still to be specified. a_i and a_i^+ are not operators in a single-particle Hilbert space H. They are operators in $H^{(A)}$ connecting two different single-particle Hilbert spaces. We shall be interested mainly in many-particle states of identical fermions. For this case, one follows the method of Jordan and Wigner (1928) and prescribes operators a_i and a_i^+ with anticommutation rules:

$$\{ a_i\ , a_j^+ \} = a_ia_j^+ + a_j^+a_i = \delta_{ij}\ ,$$

$$\{ a_i\ , a_j \} = \{ a_i^+\ , a_j^+ \} = 0\ . \qquad (15.11)$$

In terms of a_i and a_i^+ a many-particle state in occupation number space can be expressed as:

$$| n_1 n_2 \ldots\ldots n_\infty > \; = \; (a_1^+)^{n_1} (a_2^+)^{n_2} \ldots\ldots (a_\infty^+)^{n_\infty} \; |0> \quad , \tag{15.12}$$

where $|0>$ stands for a state $|n_1 n_2 \ldots\ldots n_\infty>$ where $n_1 = n_2 = \ldots\ldots = n_\infty = 0$. Because of the anticommutation rules, the order of the a_i^+ operators is important for different ordering of a_i^+ would introduce a minus sign. Applying (15.11), we find:

$$a_i \; | \ldots\ldots n_i \ldots\ldots > \; = \begin{cases} (-1)^{S_i} | \ldots\ldots n_i - 1 \ldots\ldots > & \text{if } n_i = 1, \\ 0 & \text{if } n_i = 0, \end{cases} \tag{15.13}$$

$$a_i^+ \; | \ldots\ldots n_i \ldots\ldots > \; = \begin{cases} (-1)^{S_i} | \ldots\ldots n_i + 1 \ldots\ldots > & \text{if } n_i = 0, \\ 0 & \text{if } n_i = 1, \end{cases} \tag{15.14}$$

and,

$$a_i^+ a_i | \ldots\ldots n_i \ldots\ldots > \; = \; n_i \; | \ldots\ldots n_i \ldots\ldots > \; , \tag{15.15}$$

where $S_i = n_1 + n_2 + \ldots\ldots + n_{i-1}$ is the number of occupied states before the ith state.

The many-body Schrodinger equation (1.1) assumes the following form in the occupation number space representation:

$$\hat{H} \; | \; \Psi > \; = \; E \; | \; \Psi > \; , \tag{15.16}$$

where,

$$\hat{H} \; = \; \sum \; a_i^+ <i| \; T \; |j> a_j \; + \frac{1}{2} \; \sum \; a_i^+ a_j^+ <ij| \; V \; |mn> \; a_n a_m \; , \tag{15.17}$$

with,

$$<i \; | \; T \; | \; j> \; = \; \int d^3x \; \psi_i^+(\vec{x}) \; \left(- \frac{\hbar^2}{2m} \nabla^2\right) \psi_j(\vec{x}) \; ,$$

and

$$<ij \; | \; V \; | \; mn> \; = \; \int d^3x d^3y \; \psi_i^+(\vec{x}) \psi_j^+(\vec{y}) v(\vec{x}-\vec{y}) \psi_m(\vec{x}) \psi_n(\vec{y}) \, . \tag{15.18}$$

In (15.17) we have dropped from (1.1) the background potentials V. We shall verify this result for the Hartree-Fock case. The Hartree-Fock many-body state is given by:

$$|\Psi> \; = \; a_1^+ a_2^+ \ldots\ldots a_A^+ \; |0> \; , \tag{15.19}$$

and then,

$$a_i^+ a_j \; |\Psi> \; = \; \delta_{ij} \; n_i |\Psi> \; + \; |\Psi'> \; ,$$

$$a_i^+ a_j^+ a_n a_m \; |\Psi> \; = \; n_i n_j (\delta_{im} \delta_{jn} \; - \; \delta_{in} \delta_{jm}) |\Psi> \; + \; |\Psi''> \; , \tag{15.20}$$

where $|\Psi'\rangle$ and $|\Psi''\rangle$ are orthogonal to $|\Psi\rangle$. Thus,

$$\langle\Psi|\hat{H}|\Psi\rangle = \sum_i \int d^3x\, \psi_i^+(\vec{x})\left(-\frac{\hbar^2}{2m}\nabla^2\right)\psi_i(\vec{x}) + $$
$$+ \frac{1}{2}\sum_{i,j}\int d^3x d^3y\, \psi_i^+(\vec{x})\psi_j^+(\vec{y})v(\vec{x}-\vec{y})[\psi_i(\vec{x})\psi_j(\vec{y}) - \psi_j(\vec{x})\psi_i(\vec{y})\}$$

(15.21)

which is the desired Hartree-Fock result. It should be noted that the ordering of the final two annihilation operators in $\hat{H}$ is opposite to that of the last two single-particle wave functions in the matrix elements of v.

It is often convenient to define field operators $\hat{\psi}$ and $\hat{\psi}^+$ which are linear combinations of the annihilation and creation operators in the occupation number space:

$$\hat{\psi}(\vec{x}) = \sum_k \psi_k(\vec{x})\, a_k ,$$
$$\hat{\psi}^+(\vec{x}) = \sum_k \psi_k^+(\vec{x})\, a_k^+ , \qquad (15.22)$$

where the coefficients are the single-particle wave functions, and the summation is over the complete set of single-particle quantum numbers, such as momentum, spin and isospin components. The field operators satisfy the following commutation relations:

$$\{\hat{\psi}(\vec{x}), \hat{\psi}^+(\vec{x}')\} = \sum_k \psi_k(\vec{x})\psi_k^+(\vec{x}') = \delta(\vec{x}-\vec{x}') ,$$
$$\{\hat{\psi}(\vec{x}), \hat{\psi}(\vec{x}')\} = \{\hat{\psi}^+(\vec{x}), \hat{\psi}^+(\vec{x}')\} = 0 , \qquad (15.23)$$

where the completeness of the single-particle wave functions is made use of.

The Hamiltonian operator $\hat{H}$ can be rewritten in terms of these field operators as follows:

$$\hat{H} = \int d^3x\, \hat{\psi}^+(\vec{x})\left(-\frac{\hbar^2}{2m}\nabla^2\right)\hat{\psi}(\vec{x}) + \frac{1}{2}\iint d^3x d^3y\, \hat{\psi}^+(\vec{x})\hat{\psi}^+(\vec{y})v(\vec{x}-\vec{y})\hat{\psi}(\vec{y})\hat{\psi}(\vec{x}).$$

(15.24)

The extension of this formalism to any other operators in the occupation number representation is clear. In general, any one-particle operator may be written in a form analogous to the kinetic energy term of $\hat{H}$, and a two-particle operator like the potential term of H. For example, the total number operator assumes the form:

$$\hat{N} = \int d^3x\, \hat{\psi}^{+}(\vec{x})\hat{\psi}(\vec{x}) = \sum_k a_k^{+} a_k = \sum_k n_k \,. \tag{15.25}$$

Field theory is formulated with field operators which are time-dependent. This is accomplished by the following unitary transformation:

$$\begin{aligned}\hat{\psi}(x) = \hat{\psi}(\vec{x},t) &= e^{i\hat{H}t/\hbar}\, \hat{\psi}(\vec{x})\, e^{-i\hat{H}t/\hbar} \\ &= \hat{\psi}(\vec{x}) + \frac{(it/\hbar)}{1!}[\hat{H},\hat{\psi}(\vec{x})] + \frac{(it/\hbar)^2}{2!}[\hat{H},[\hat{H},\hat{\psi}(\vec{x})]] + \\ &\quad + \frac{(it/\hbar)^3}{3!}[\hat{H},[\hat{H},[\hat{H},\hat{\psi}(\vec{x})]]] + \ldots \end{aligned} \tag{15.26}$$

Description of quantum mechanical system in terms of these operators is said to be in the Heisenberg representation, for which the state vectors forming the Hilbert space are time-independent. The Heisenberg representation thus offers a description of the dynamical system entirely in terms of field operators without the need to refer to the state vectors. The incorporation of the time variable together with the spatial variables in the description of the field is indispensable to a relativistic formulation of dynamics. We shall designate field operators in the Heisenberg representation by displaying their arguments $x = (\vec{x},t)$. The equal-time commutation relations of these fields are similar to those of (15.23):

$$\begin{aligned}\{\hat{\psi}(\vec{x},t),\hat{\psi}^{+}(\vec{x}',t)\} &= e^{i\hat{H}t/\hbar}\,\{\hat{\psi}(\vec{x}),\, \hat{\psi}(\vec{x}')\}\, e^{-i\hat{H}t/\hbar} = \delta(\vec{x}-\vec{x}')\,, \\ \{\hat{\psi}(\vec{x},t),\, \hat{\psi}(\vec{x}',t)\} &= \{\hat{\psi}^{+}(\vec{x},t),\hat{\psi}^{+}(\vec{x}',t)\} = 0\,. \end{aligned} \tag{15.27}$$

The unequal-time commutation relations of these fields cannot be specified so easily, since it necessitates in general the solution of the dynamical problem before they are evaluated. The field operator satisfies the following equation of motion:

$$i\hbar\, \partial_t \hat{\psi}(\vec{x},t) = [\, \hat{\psi}(\vec{x},t),\, \hat{H}\,]\,. \tag{15.28}$$

The most systematic formulation of quantum field theory is by means of a Lagrangian:

$$L = \int d^3x \; \mathcal{L}(\hat{\psi}, \partial_\mu \hat{\psi}) \; , \tag{15.29}$$

where $\mathcal{L}$ is called the Lagrangian density and is a function of the fields $\hat{\psi}$ and their derivatives $\partial_\mu \psi$, which are treated as independent variables. It is then required that the action integral:

$$I = \int_{\Omega_4} d^4x \;\; \mathcal{L}(\hat{\psi}, \partial_\mu \hat{\psi}) \; , \tag{15.30}$$

be stationary for arbitrary variations of the field quantities, which vanish at the boundary of the 4-dimensional space-time volume Ω_4. This requirement imposes the following Euler equations:

$$\frac{\partial \mathcal{L}}{\partial \hat{\psi}} - \partial_\nu \frac{\partial \mathcal{L}}{\partial(\partial_\nu \hat{\psi})} = 0 \; . \tag{15.31}$$

These are identified with the field equations by a suitable choice of $\mathcal{L}$. However, two Lagrangian densities differing by a four-divergence yield the same action integral and result in the same field equations.

A canonical momentum $\hat{\pi}$ conjugate to $\hat{\psi}$ may be defined by:

$$\hat{\pi} = \frac{\partial \mathcal{L}}{\partial(\partial_t \hat{\psi})} \; , \tag{15.32}$$

which according to the canonical quantization procedure obeys the following equal-time anticommutation rule:

$$\{\hat{\psi}(x), \; \hat{\pi}(x)\} = i\hbar \; . \tag{15.33}$$

For fields which satisfy the equations of motion (15.31), the following (energy-momentum) stress tensor can be constructed:

$$\hat{T}_{\mu\nu} = \mathcal{L}\delta_{\mu\nu} - \frac{\partial \mathcal{L}}{\partial(\partial_\mu \hat{\psi})} \partial_\nu \hat{\psi} \; , \tag{15.34}$$

which is conserved in the sense that $\partial_\mu \hat{T}_{\mu\nu} = 0$. This tensor is not yet put into a symmetric form but may be symmetrized readily (see, for example, Wentzel 1949). The energy-momentum four-vector $\hat{P}_\mu$ is obtained from:

$$\hat{P}_\mu = \int d^3x \frac{1}{i}\hat{T}_{4\mu} , \tag{15.35}$$

from which the total energy, or the Hamiltonian, is obtained:

$$\hat{H} = \hat{P}_\theta = -\int d^3x \, \hat{T}_{44} . \tag{15.36}$$

The relativistic model of nuclear matter may be represented by the following Lagrangian density:

$$L = L^o_N + L^o_\sigma + L^o_\omega + L^I_\sigma + L^I_\omega , \tag{15.37}$$

where

$$L^o_N = -\frac{1}{2}\hbar c \, \hat{\psi}^+\gamma_4(\gamma_\mu\partial_\mu + \frac{mc}{\hbar})\hat{\psi} - \frac{1}{2}\hbar c(-\partial_\mu\hat{\psi}^+\gamma_4\gamma_\mu + \frac{mc}{\hbar}\hat{\psi}^+\gamma_4)\hat{\psi} ,$$

$$L^o_\sigma = -\frac{1}{2}c^2 \{\partial_\mu\hat{\phi}\partial_\mu\hat{\phi} + (m_\sigma c/\hbar)^2\hat{\phi}^2\}$$

$$L^o_\omega = -\frac{1}{4}\hat{F}_{\mu\nu}\hat{F}_{\mu\nu} - \frac{1}{2}(m_\omega c/\hbar)^2 \hat{V}_\mu\hat{V}_\mu$$

$$L^I_\sigma = g_\sigma \hat{\psi}^+\gamma_4\hat{\phi}\hat{\psi}$$

$$L^I_\omega = ig_\omega \hat{\psi}^+\gamma_4\gamma_\mu\hat{V}_\mu\hat{\psi}$$

$$\hat{F}_{\mu\nu} = \partial_\mu\hat{V}_\nu - \partial_\nu\hat{V}_\mu , \qquad \partial_\mu\hat{V}_\mu = 0 . \tag{15.38}$$

We shall henceforth set $\hbar = c = 1$ in this section. The equations of motion are therefore:

$$(\gamma_\mu\partial_\mu + m)\hat{\psi} = ig_\omega \gamma_\mu\hat{V}_\mu\hat{\psi} + g_\sigma\hat{\phi}\hat{\psi} ,$$

$$(\partial_\mu\partial_\mu - m^2_\sigma)\hat{\phi} = -g_\sigma \hat{\psi}^+\gamma_4\hat{\psi} ,$$

$$(\partial_\mu\partial_\mu - m^2_\omega)\hat{V}_\nu = -ig_\omega \hat{\psi}^+\gamma_4\gamma_\nu\hat{\psi} . \tag{15.39}$$

The solutions of the meson fields from the equations of motion are given by:

$$\hat{\phi}(x) = g_\sigma \int d^3x \, D_\sigma(x-x')\hat{\psi}^+(x')\gamma_4\hat{\psi}(x') , \tag{15.40}$$

$$\hat{V}_\mu(x) = ig_\omega\int d^3x \, D_\omega(x-x')\hat{\psi}^+(x')\gamma_4\gamma_\mu\hat{\psi}(x'), \tag{15.41}$$

where D_σ and D_ω are the retarded Green's functions for the Klein-Gordon equations:

$$(\partial_\mu \partial_\mu - m_a^2)\, D_a(x-x') = -\delta(x-x'), \qquad (a = \sigma, \omega) \tag{15.42}$$

and,

$$D_a(x-x') = 0 \ , \ \text{for } x_0 - x_0' < 0 \ . \tag{15.42a}$$

$D_a(x-x')$ are evaluated analogous to $g(\vec{r}-\vec{r}')$ of (11.15) from their Fourier transforms:

$$D_a(\vec{x},t) = (2\pi)^{-4} \int \frac{dk_0 d^3k \, e^{-ik_0 t} \, e^{i\vec{k}\cdot\vec{x}}}{\vec{k}^2 + m_a^2 - (k_0 + i\varepsilon)^2} \tag{15.43}$$

where the $(i\varepsilon)$ is added to obtain the retarded solutions. When k_0-integration is performed by means of coutour integral, the $(i\varepsilon)$ quantity will displace the poles below the real k_0-axis, and the integral evaluates to zero for $t < 0$. For $t > 0$, we have:

$$\begin{aligned} D_a(\vec{x},t) &= -(2\pi)^{-3} \int \frac{d^3k \, e^{i\vec{k}\cdot\vec{x}} \sin(\sqrt{\vec{k}^2+m_a^2}\, t)}{\sqrt{\vec{k}^2+m_a^2}} \\ &= -\frac{1}{2\pi^2} \frac{1}{r} \int_0^\infty k dk \frac{\sin(kr) \sin(\sqrt{\vec{k}^2+m_a^2}\, t)}{\sqrt{\vec{k}^2+m_a^2}} \\ &= \frac{1}{4\pi} \frac{1}{r} \frac{\partial}{\partial r} \left\{ \frac{1}{\pi} \int_0^\infty \frac{dk}{\sqrt{\vec{k}^2+m_a^2}} \cos(kr) \sin(\sqrt{\vec{k}^2+m_a^2}\, t) \right\}, \end{aligned} \tag{15.44}$$

where $r = |\vec{x}|$, and the quantity in the brackets may be identified with the Bessel functions (see Schweber 1961):

$$\{ \qquad \} = \begin{cases} J_0(m_a\sqrt{t^2-r^2}) & \text{for } t > r \ , \\ 0 & \text{for } -r < t < r \ , \\ -J_0(m_a\sqrt{t^2-r^2}) & \text{fro } t < -r \ . \end{cases} \tag{15.45}$$

Hence, D_a may be given explicitly as follows:

$$D_a(\vec{x},t) = \Theta(t) \left\{ -\frac{1}{2\pi} \delta(t^2-r^2) - \frac{1}{2} m_a^2 \,\Theta(t^2-r^2) \frac{J_1(m_a\sqrt{t^2-r^2})}{m_a\sqrt{t^2-r^2}} \right. \tag{15.46}$$

where $\Theta(z)$ is the step-function of (4.33), and the δ-function is obtained from the differentiation of (15.45), which for $|t| \to r$ is given by:

$$\{ \qquad \} = \Theta(t-r) - \Theta(-t-r) \ . \tag{15.47}$$

D_a are written in a manifestly covariant form as;

$$D_a(x) = \Theta(x_0)\{-\frac{1}{2\pi}\delta(-x_\mu x_\mu) - \frac{1}{2}m_a^2\,\Theta(-x_\mu x_\mu)\frac{J_1(m_a\sqrt{-x_\mu x_\mu})}{m_a\sqrt{-x_\mu x_\mu}}\}. \qquad (15.48)$$

The Hamiltonian $\hat{H}$ (15.36) can now be expressed entirely in terms of the nucleon fields;

$$\hat{H} = \hat{H}_0 + \hat{H}_I , \qquad (15.49)$$

where

$$\hat{H}_0 = \int d^3x\, \hat{\psi}^+(x)\gamma_4(\gamma_\mu\partial_\mu + m)\hat{\psi}(x) ,$$

$$\hat{H}_I = \frac{1}{2}\int d^3x_1\int d^4x_2\, \{-g_\sigma^2\, \hat{\psi}^+(x_1)\gamma_4(1)\, \hat{\psi}^+(x_2)\gamma_4(2)\, D_\sigma(x_1-x_2)\hat{\psi}(x_2)\hat{\psi}(x_1) +$$

$$+ g_\omega^2\, \hat{\psi}^+(x_1)\gamma_4(1)\gamma_\mu(1)\hat{\psi}^+(x_2)\gamma_4(2)\gamma_\mu(2)\, D_\omega(x_1-x_2)\hat{\psi}(x_2)\hat{\psi}(x_1)\}. \qquad (15.50)$$

The nucleon fields may be expanded into a complete set of Dirac spinors:

$$\hat{\psi}(x) = \hat{\psi}(\vec{x},t) = \Omega^{-1/2}\sum_{\vec{p}}\sum_{r=1}^{4} a_{\vec{p}}^{(r)}(t)\; u^{(r)}(\vec{p})\; e^{i\vec{p}\cdot\vec{x}} , \qquad (15.51)$$

where $a_{\vec{p}}^{(r)}(t)$ are time-dependent annihilators. $\hat{\psi}^+(x)$ are expanded similarly but in terms of the creation operators, $a_{\vec{p}}^{(r)+}(t)$. These operators satisfy the same anticommutation relations as a_i (15.11) at equal times:

$$\{ a_{\vec{p}}^{(r)}(t),\, a_{\vec{p}'}^{(r')+}(t)\} = \delta_{rr'}\,\delta_{\vec{p}\vec{p}'} ,$$

$$\{ a_{\vec{p}}^{(r)}(t),\, a_{\vec{p}'}^{(r')}(t) \} = \{ a_{\vec{p}}^{(r)+}(t),\, a_{\vec{p}'}^{(r')+}(t)\} = 0 . \qquad (15.52)$$

The time dependences of $a_{\vec{p}}^{(r)}(t)$ and $a_{\vec{p}}^{(r)+}(t)$ are to be deduced from the Heisenberg equation of motion (15.28) by commuting them with $\hat{H}$. With $\hat{H}$ given by (15.49), this operation is impossible since $\hat{\psi}(x_2)$ in $\hat{H}_I$ belongs to different times and the commutation relation is not known. Taking the perturbative approach, the time dependence of the operators is established by replacing $\hat{H}$ by $\hat{H}_0$, which contains nucleon fields at a single time. $\hat{H}_0$ may be reduced as follows;

$$\hat{H}_0 = \Omega^{-1}\int d^3x \sum_{\vec{p},\vec{p}'} \sum_{r,r'} a_{\vec{p}}^{(r)+}(t)u^{(r)+}(\vec{p})\, e^{-i\vec{p}\cdot\vec{x}}$$

$$(\vec{\alpha}\cdot\vec{p}'+\beta m)\; a_{\vec{p}'}^{(r')}(t)u^{(r')}(\vec{p}')\, e^{i\vec{p}'\cdot\vec{x}}$$

$$= \sum_{\vec{p},\vec{p}'} \sum_{r,r'} \delta_{\vec{p}\vec{p}'}\, E'\, a_{\vec{p}}^{(r)+}(t)\; a_{\vec{p}'}^{(r')}(t)\; u^{(r)+}(\vec{p})u^{(r')}(\vec{p}')$$

$$= \sum_{\vec{p}} \{ \sum_{r=1,2} p_0\, a_{\vec{p}}^{(r)+}a_{\vec{p}}^{(r)} - \sum_{r=3,4} p_0\, a_{\vec{p}}^{(r)+}a_{\vec{p}}^{(r)} \} \,, \qquad (15.53)$$

where $p_0 = \sqrt{\vec{p}^{\,2}+m^2}$. The approximate time dependence of the operators is computed from:

$$\partial_t\, a_{\vec{p}}^{(r)}(t) = -i\left[\, a_{\vec{p}}^{(r)}(t),\, H_0 \,\right] = \mp\, ip_0\, a_{\vec{p}}^{(r)}(t) \quad \text{for } r = \{{1,2 \atop 3,4} \,,$$

$$\partial_t\, a_{\vec{p}}^{(r)+}(t) = -i\left[\, a_{\vec{p}}^{(r)+}(t), H_0 \,\right] = \pm\, ip_0\, a_{\vec{p}}^{(r)+}(t) \quad \text{for } r = \{{1,2 \atop 3,4} \,.$$

$$(15.54)$$

In deducing (15.54) we have made use of the relation that:

$$\left[\, AB,\, C \,\right] = A\{\, B,C \,\} - \{\, A,C\}\, B \;, \qquad (15.55)$$

where A, B, and C are operators. Writing out the time-dependence explicitly, we have:

$$a_{\vec{p}}^{(r)}(t) = a_{\vec{p}}^{(r)}(0)\; e^{\mp ip_0 t} \qquad \text{for } r = \{{1,2 \atop 3,4} \;,$$

$$a_{\vec{p}}^{(r)+}(t) = a_{\vec{p}}^{(r)+}(0)\; e^{\pm ip_0 t} \qquad \text{for } r = \{{1,2 \atop 3,4} \;. \qquad (15.56)$$

Subsequently, we shall write $a_{\vec{p}}^{(r)}(0) = a_{\vec{p}}^{(r)}$ and $a_{\vec{p}}^{(r)+}(0) = a_{\vec{p}}^{(r)+}$.

In the Dirac hole theory, all negative energy states are assumed completely filled under normal conditions, for if not, all particles in the positive energy states can and will make transitions to these states. Consequently, the annihilation of a negative energy particle of momentum $-\vec{p}$ and spin-down is interpreted as the creation of an antiparticle with momentum $+\vec{p}$ and spin-up. To achieve this interpretation, one redefines:

$$b_{\vec{p}}^{(1)+}(t) = -\, a_{-\vec{p}}^{(4)}(t) \;, \qquad v^{(1)}(\vec{p}) = -\, u^{(4)}(-\vec{p}) \;,$$

$$b_{\vec{p}}^{(2)+}(t) = a_{-\vec{p}}^{(3)}(t) \;, \qquad v^{(2)}(\vec{p}) = u^{(3)}(-\vec{p}) \;. \qquad (15.57)$$

The b^+ and b operators satisfy the same equal-time commutation relations as a^+ and a. The expansions for $\hat{\psi}$ and $\hat{\psi}^+$ now take the following forms:

$$\hat{\psi}(\vec{x},t) = \Omega^{-1/2}\sum_{\vec{p}}\sum_{s=1,2}\left(a_{\vec{p}}^{(s)} u^{(s)}(\vec{p})e^{ip\cdot x} + b_{\vec{p}}^{(s)+} v^{(s)}(\vec{p})e^{-ip\cdot x}\right),$$

$$\hat{\psi}^+(\vec{x},t) = \Omega^{-1/2}\sum_{\vec{p}}\sum_{s=1,2}\left(b_{\vec{p}}^{(s)} v^{(s)+}(\vec{p})e^{ip\cdot x} + a_{\vec{p}}^{(s)+} v^{(s)+}(\vec{p})e^{-ip\cdot x}\right),$$

(15.58)

where $p\cdot x = \vec{p}\cdot\vec{x} - p_0 t$. $\hat{H}_0$ may be rewritten as:

$$\hat{H}_0 = \sum_{\vec{p}}\sum_{s} p_0 \left(a_{\vec{p}}^{(s)+} a_{\vec{p}}^{(s)} + b_{\vec{p}}^{(s)+} b_{\vec{p}}^{(s)}\right) - \sum_{\vec{p}}\sum_{s} p_0 \,, \qquad (15.59)$$

where the last term corresponding to an infinite energy of the filled negative energy states is to be dropped from dynamical considerations.

Applying the field theory formalism to the nuclear matter problem, $a_{\vec{p}}$ and $a_{\vec{p}}^+$ are respectively the annihilation and creation operator for nucleons, and $b_{\vec{p}}$ and $b_{\vec{p}}^+$ are respectively the annihilation and creation operator for antinucleons. We shall omit terms proportional to $b_{\vec{p}}$ and $b_{\vec{p}}^+$ in the following discussion, since in the lowest order approximation they will not play a role. In general, $\hat{\psi}$ and $\hat{\psi}^+$ need not be expanded into plane wave solutions of the Dirac equation, and may be expanded into any form of spinor solutions $f_\alpha(\vec{x})$ of the Dirac equation. We shall in fact write $\hat{\psi}$, $\hat{\psi}^+$ in such forms: (suppressing b,b^+ terms)

$$\hat{\psi} = \sum_{\alpha} f_\alpha(\vec{x}) a_\alpha e^{-iE_\alpha t} \,,$$

$$\hat{\psi}^+ = \sum_{\alpha} f_\alpha^+(\vec{x}) a_\alpha^+ e^{iE_\alpha t} \,, \qquad (15.60)$$

where E_α are the eigenvalues of the Dirac equation. (Note that the Dirac equation needs not be the free equation, but it should be linear in the spinor wave function.)

Inserting $\hat{\psi}$ and $\hat{\psi}^+$ into H_I of (15.50), we find:

$$H_I = \frac{1}{2}\sum_{\alpha,\alpha,\beta,\beta'}\int d^3x_1 d^3x_2 f_{\alpha'}^+(\vec{x}_1) f_{\beta'}^+(\vec{x}_2)\{v_\sigma(|\vec{x}_1-\vec{x}_2|)+v_\omega(|\vec{x}_1-\vec{x}_2|)\}_{\alpha,\alpha'} \times$$

$$\times f_\beta(\vec{x}_2) f_\alpha(\vec{x}_1) a_{\alpha'}^+ a_{\beta'}^+ a_\beta a_\alpha \qquad (15.61)$$

where,

$$\{v_\sigma\}_{\alpha,\alpha'} = -g_\sigma^2 \gamma_4(1)\gamma_4(2)\int dt_2 e^{iE_{\alpha'}t_1} e^{iE_{\beta'}t_2} D_\sigma(x_1-x_2) e^{-iE_\beta t_2} e^{-iE_\alpha t_1} ,$$

and,

$$\{v_\omega\}_{\alpha,\alpha'} = g_\omega^2 \gamma_4(1)\gamma_\mu(1)\gamma_4(2)\gamma_\mu(2)\int dt_2 e^{iE_{\alpha'}t_1} e^{iE_{\beta'}t_2} D_\omega(x_1-x_2) \times$$
$$\times e^{-iE_\beta t_2} e^{-iE_\alpha t_1} . \qquad (15.62)$$

These expressions are evaluated most easily using a form of D_a given by (15.44). For example, the t_2-integrals of (15.62) are given by:

$$e^{i(E_{\alpha'}-E_\alpha+E_{\beta'}-E_\beta)t_1} \int_\infty^{t_1} dt_2\, e^{i(E_\beta-E_{\beta'})(t_1-t_2)} \times$$
$$\times \; -(2\pi)^{-3}\int d^3p \frac{e^{i\vec{p}\cdot\vec{x}}}{\sqrt{\vec{p}^2+m_a^2}} \sin\left(\sqrt{\vec{p}^2+m_a^2}\,(t_1-t_2)\right)$$
$$= (2\pi)^{-3}\int d^3p \frac{e^{i\vec{p}\cdot\vec{x}}}{\sqrt{\vec{p}^2+m_a^2}} \int_0^\infty dt\, e^{i(E_\beta-E_{\beta'})t} \sin\left(\sqrt{\vec{p}^2+m_a^2}\, t\right)$$
$$= (2\pi)^{-3}\int d^3p \frac{e^{i\vec{p}\cdot\vec{x}}}{\vec{p}^2+m_a^2-(E_\beta-E_{\beta'})^2} = (2\pi)^{-2} \frac{1}{ir}\int_{-\infty}^{\infty} \frac{p\,dp\, e^{ipr}}{p^2+\left(m_a^2-(E_\beta-E_{\beta'})^2\right)}$$
$$= \frac{1}{4\pi}\frac{1}{r} \exp\{-\sqrt{m_a^2-(E_\beta-E_{\beta'})^2}\; r\} , \qquad (15.63)$$

where energy conservation, $E_{\alpha'}+E_{\beta'}=E_\alpha+E_\beta$ is made used of. Hence,

$$\{v_\sigma\}_{\alpha,\alpha'} = -(g_\sigma^2/4\pi)\beta(1)\beta(2)r^{-1}\exp\{-\sqrt{m_\sigma^2-(E_\alpha-E_{\alpha'})^2}\; r\} ,$$
$$\{v_\omega\}_{\alpha,\alpha'} = (g_\omega^2/4\pi)\{I - \vec{\alpha}(1)\cdot\vec{\alpha}(2)\}\; r^{-1}\exp\{-\sqrt{m_\omega^2-(E_\alpha-E_{\alpha'})^2}\; r\} . \qquad (15.64)$$

These are the potential functions employed in (14.62).

The results of Section 14 will be obtained if the total energy of the system is evaluated by:

$$E = \langle\Psi|\, \hat{H}\, |\Psi\rangle , \qquad (15.65)$$

where the ground state vector is given by (15.19) and f_α by $u(\vec{p})e^{i\vec{p}\cdot\vec{x}}$ of (14.55).

There is an alternate approximate treatment of the nuclear matter problem. This is by means of the mean field approach. It consists in replacing the meson fields by their ground state expectation values, which are then just classical fields:

$$\hat{\phi} \rightarrow \phi = \langle\Psi|\ \hat{\phi}\ |\Psi\rangle ,$$

and,

$$\hat{V}_\mu \rightarrow V_\mu = \langle\Psi|\ \hat{V}_\mu\ |\Psi\rangle . \tag{15.66}$$

In so doing, the equation of motion for the nucleon fields is linearized and may be solved readily. For example, in the nuclear matter problem, the ground state is expected to be uniform and isotropic, and hence the ground state expectation value of the vector components of $\hat{V}_\mu$ should vanish, $\vec{V} = 0$, and ϕ and $V_4 = iV_0$ to be independent of $\vec{x}$ and t. From their equations of motion, it follows that:

$$\phi = (g_\sigma/m_\sigma^2)\ \langle\Psi|\ \hat{\psi}^+\beta\hat{\psi}\ |\Psi\rangle \equiv (g_\sigma/m_\sigma^2)\ n_s , \tag{15.67}$$

$$V_0 = (g_\omega/m_\omega^2)\ \langle\Psi|\ \hat{\psi}^+\ \hat{\psi}\ |\Psi\rangle \equiv (g_\omega/m_\omega^2)\ n_B . \tag{15.68}$$

In (15.68), the expectation value of $\hat{\psi}^+\hat{\psi}$ in a uniform ground state is given by the nucleon number density n_B because $\int d^3x\hat{\psi}^+\hat{\psi}$ is in fact the number operator (15.25), which gives the total number of nucleons in the system. n_s is to be determined below.

The equation of motion for the nucleon field is simplified to read:

$$\{\ \gamma_\mu\partial_\mu + m - g_\sigma\phi + g_\omega\gamma_4V_0\}\hat{\psi} = 0 . \tag{15.69}$$

Expanding $\hat{\psi}$ and $\hat{\psi}^+$ as in (15.60), we get:

$$\hat{\psi}(\vec{x},t) = \sum_{r=1,2} u^{(r)}(\vec{p})\ a_{\vec{p}}\ e^{i(\vec{p}\cdot\vec{x} - E_pt)} ,$$

$$\hat{\psi}^+(\vec{x},t) = \sum_{r=1,2} u^{(r)+}(\vec{p})\ a^+_{\vec{p}}\ e^{-i(\vec{p}\cdot\vec{x} - E_pt)} , \tag{15.70}$$

where $u^{(r)}(\vec{p})$ and E_p are solutions of:

$$\{\ \vec{\alpha}\cdot\vec{p} + (m - g_\sigma\phi) + g_\omega V_0\ \}u^{(r)}(\vec{p}) = E_p\ u^{(r)}(\vec{p}) . \tag{15.71}$$

Writing $(m-g_\sigma\phi) = m^*$, the effective mass, the Dirac spinors are obtained from the solution of:

$$(\ \vec{\alpha}\cdot\vec{p} + m^*)\ u^{(r)}(\vec{p}) = (E_p - g_\omega V_0)u^{(r)}(\vec{p}) = (\vec{p}^2 + m^{*2})^{1/2}u^{(r)}(\vec{p}) , \tag{15.72}$$

where the last line is obtained by "squaring" both sides of the equation. The adjoint equation of (15.72) is given by:

$$u^{(r)+}(\vec{p})\ (\vec{\alpha}\cdot\vec{p} + \beta m^{*}) = (\vec{p}^{\,2}+m^{*2})^{1/2}\ u^{(r)+}(\vec{p})\ . \qquad (15.73)$$

From (15.72) and (15.73), it can be deduced that;

$$u^{(r)+}(\vec{p})\ u^{(r)}(\vec{p})\ m^{*} = u^{(r)+}(\vec{p})\ \beta\ u^{(r)}(\vec{p})\ (\vec{p}^{\,2}+m^{*2})^{1/2}. \qquad (15.74)$$

Let us evaluate now n_s:

$$\begin{aligned} n_s &= \langle\Psi|\ \hat{\bar\psi}\ \beta\hat{\psi}\ |\Psi\rangle = \sum_{\vec{p},r} u^{(r)+}(\vec{p})\ \beta\ u^{(r)}(\vec{p}) \\ &= \sum_{\vec{p},r} u^{(r)+}(\vec{p})\ u^{(r)}(\vec{p})\ \frac{m^{*}}{(\vec{p}^{\,2}+m^{*2})^{1/2}} = \frac{\gamma}{(2\pi)^3}\int_0^{k_F} \frac{d^3p\ m^{*}}{(\vec{p}^{\,2}+m^{*2})^{1/2}}\ . \end{aligned} \qquad (15.75)$$

Comparing these expressions with those of Section 14, it is clear that (15.71) is identical to (14.34) with $m^{*} = m_H$, and the expression for n_s is the same as (14.39). Hence, the present mean field approximation yeilds the same results as the Hartree approximation of Section 14.

The energy density of the system is evaluated from the Hamiltonian:

$$\begin{aligned} \frac{\langle\Psi|\hat{H}|\Psi\rangle}{\Omega} &= \frac{1}{\Omega}\int d^3x\ \langle\Psi|\ \hat{\bar\psi}\ \{-i\vec{\alpha}\cdot\vec{\nabla} + \beta(m-g_\sigma\phi) + g_\omega V_0\}\ \hat{\psi}\ + \frac{1}{2}m_\sigma^2\ \phi^2 + \frac{1}{2}m_\omega^2 V_0^2 \\ &= \sum_{\vec{p},r} (g_\omega V_0 + \sqrt{\vec{p}^{\,2}+m^{*2}}) + \frac{1}{2}m_\sigma^2\phi^2 - \frac{1}{2}m_\omega^2 V_0^2 \\ &= \gamma(2\pi)^{-3}\int d^3p\ \sqrt{\vec{p}^{\,2}+m^{*2}} + \frac{1}{2}(g_\sigma^2/m_\sigma^2)n_s^2 + \frac{1}{2}(g_\omega^2/m_\omega^2)n_B^2. \end{aligned} \qquad (15.76)$$

For a uniform fluid at rest the pressure is given by the stress tensor as:

$$\begin{aligned} P &= \langle\Psi|\ \frac{1}{3}\hat{T}_{ii}\ |\Psi\rangle = \langle\Psi|\frac{1}{3}\ \hat{\psi}^{+}(-i\vec{\alpha}\cdot\vec{\nabla}) - \frac{1}{2}m_\sigma^2\phi^2 + \frac{1}{2}m_\omega^2 V_0^2\ |\Psi\rangle \\ &= \frac{1}{3}\frac{\gamma}{(2\pi)^3}\int_0^{k_F}\frac{d^3p\ p^2}{(\vec{p}^{\,2}+m^{*2})^{1/2}} - \frac{1}{2}(g_\sigma^2/m_\sigma^2)n_s^2 + \frac{1}{2}(g_\omega^2/m_\omega^2)n_B^2\ . \end{aligned} \qquad (15.77)$$

The mean field results are just those of the Walecka model quoted in Section 14. The Walecka model was originally presented in the mean field approach (Walecka 1974). The extension of the mean field approach to include Hartree-Fock exchange effects had been carried out by Chin (1977). We shall not include such a discussion here.

The purpose of the mean field approximations of the meson fields is to decouple these fields as dynamical variables from the system. They contribute merely certain effective interactions to the nucleon Lagrangian. These effective interactions would have the same Lorentz transformation properties as the fields, and would contribute to the respective Γ_a terms listed in (14.22). As effective interactions, the mean fields are also density dependent. For example, V_0 is proportional to n_B while ϕ to n_s. As we have discussed in Section 14, since the Walecka model predicts a nuclear compressibility K which is far too large, it suggests that the density dependence of the derived effective interaction may not be adequate. To modify the situation, one possible way of doing it is by adding non-linear self-interaction terms of meson fields into the Lagrangian. An interesting modification has been suggested by the non-linear σ-model, for which L_σ^o of (15.38) is replaced by:

$$L_\sigma^o = -\frac{1}{2}(\partial_\mu\hat{\phi}\partial_\mu\hat{\phi}) - U(\hat{\phi}) ,$$

where,

$$U(\hat{\phi}) = \frac{1}{2} m_\sigma^2 \hat{\phi}^2 + \frac{1}{3} b \hat{\phi}^3 + \frac{1}{4} c \hat{\phi}^4 , \tag{15.78}$$

where b and c are adjustable parameters. In the mean field approximation ϕ is to be solved from:

$$m_\sigma^2\phi + b\phi^2 + c\phi^3 = g_\sigma n_s . \tag{15.79}$$

Boguta and Bodmer (1977) have investigated this model and they find that by choosing:

$$g_\omega^2/m_\omega^2 = 1/m^2 , \quad g_\sigma^2/m_\sigma^2 = 8/m^2 , \quad m_\sigma = 250 \text{ MeV} ,$$
$$b = 0.445\, g_\sigma m , \qquad c = 9.465\, g_\sigma^4 , \tag{15.80}$$

the nuclear compressibility is reduced to K = 150 MeV. However, it is difficult to trace the origin of the nonlinear terms. They have been attributed to three- and four-body forces, although a direct connection is difficult to establish. Banerjee, Glendenning and Gyulassy (1981) applied the nonlinear model to study pion condensation in nuclear matter. They investigated several alternative solutions to the nuclear matter problem and they are:

model	$g_\omega m/m_\omega$	$g_\sigma m/m_\sigma$	$b/g_\sigma^3 m$	c/g_σ^4
a	11	15	0.004	0.008
b	5	9	-0.192	2.47
c	4.7	6	-0.734	6.89

All of these models could reproduce nuclear saturation properties and at the same time keeping the nuclear compressibility within acceptable values. It is clear that the non-linear terms can take over the role of the linear ϕ and V_0 terms in producing saturation and thus reducing the values of the coupling constants g_ω and g_σ. In fact Boguta and Bodmer (1977) showed that by adjusting the b and c parameters appropriately nuclear saturation could be obtained from this model without the ω-meson contribution. Thus, non-linear interactions can generate all sorts of results and must be handled carefully in a phenomenological approach. The non-linear terms should preferrably be justified independently before they are employed.

References

Banerjee, B., Glendenning, N.K. and Gyulassy, M. (1981). Nucl. Phys. A361, 326.

Boguta, J. and Bodmer, A.R. (1977). Nucl. Phys. A292, 413.

Brockmann, R. (1978). Phys. Rev. C18, 1510.

Chin, S.A. (1977). Ann. Phys. 108, 301.

Jordan, P. and Wigner, E. (1928). Z. Physik 47, 631.

Schweber, S.S. (1961). An Introduction to Relativistic Quantum Field Theory, Harper & Row, New York.

Walecka, J.D. (1974). Ann. Phys. 83, 491.

Wentzel, G. (1949). Quantum Theory of Fields, Interscience, New York.

Bibliography

Bjorken, J.D. and Drell, S.D. (1964). Relativistic Quantum Fields, McGraw-Hill, New York.

Fetter, A.L. and Walecka, J.D. (1971). Quantum Theory of Many-Particle Systems, McGraw-Hill, New York.

16. Baryonic Matter

For ultradense matter it becomes necessary to consider the presence of hyperons and other baryons such as the nucleon resonances like the Δ and N^* in addition to the nucleons in its composition. Also muons and possibly other heavy leptons will share the role of the electrons. All hyperons, nucleons and their resonances will be referred to collectively as baryons, for which a conserved quantum number, the baryon number, may be assigned. The conditions related to the appearance of hyperons are similar to those of the neutronization process. As matter density increases, the increase in the Fermi energy of the nucleons would make the appearance of more massive baryons possible. There are a large number of massive baryons known. We shall illustrate the method by concentrating on those with lower masses, which will contribute to the composition first as matter density increases. The method presented can be easily generalized to include the more massive baryons. The hyperons that we shall write out explicitly are: $\Lambda(1115.6)$, $\Sigma^+(1189.4)$, $\Sigma^o(1192.5)$, $\Sigma^-(1197.4)$, whose mass, or mc^2, in units of MeV are written in parantheses. They are stable particles. For nucleon resonances we include just the $\Delta(1232)$, which has charge components Δ^{++}, Δ^+, Δ^o, Δ^-. We shall for simplicity take the nucleon mass to be 939 MeV, the Λ mass to be 1115 MeV and the Σ mass to be 1190 MeV.

The weak interactions responsible for the production of hyperons are:

$$
\begin{aligned}
p + e &\to \Lambda + \nu \,, \\
p + e &\to \Sigma^o + \nu \,, \\
n + e &\to \Sigma^- + \nu \,,
\end{aligned}
\tag{16.1}
$$

while the transformation of the nucleons to Δ proceeds via the strong interaction. The conditions of chemical equilibrium are determined by the chemical potentials of these baryons:

$$
\begin{aligned}
\mu_{\Sigma^-} &= \mu_{\Delta^-} = \mu_n + \mu_e \,, \\
\mu_{\Sigma^o} &= \mu_{\Delta^o} = \mu_\Lambda = \mu_n \,, \\
\mu_{\Sigma^+} &= \mu_{\Delta^+} = \mu_p = \mu_n - \mu_e \,, \\
\mu_{\Delta^{++}} &= \mu_n - 2\,\mu_e \,.
\end{aligned}
\tag{16.2}
$$

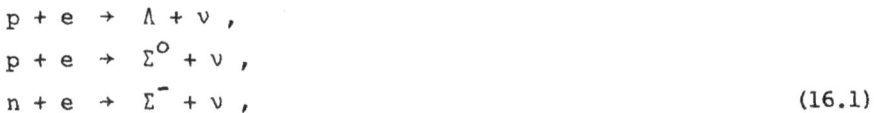

Charge neutrality of the system demands the total positively charged particles be equal to the negatively charged particles. Expressed in terms of the particles' Fermi momenta, the condition of charge neutrality is given by:

$$\sum_{a+} p_{Fa+}^{3} = \sum_{a-} p_{Fa-}^{3} , \qquad (16.3)$$

where a+ (a-) denote the positively (negatively) charged specis. For doubly charged particles, a factor of 1/2 should be introduced.

By and large, electrons are playing the role of maintaining charge neutrality. Whenever a positively charged baryon is added, an electron has to accompany it. On the other hand, when a negatively charged baryon is added, an electron may be removed from the system. Since the electrons being very light are very energetic and thus raising the energy of the system. The ground state of the system therefore favors the production of negatively charged baryons whenever it is possible. The interactions of the hyperons and Δ are very poorly known. We shall first make an order of magnitude estimate for baryonic matter by ignoring interactions (Ambartsumyan and Saakyan 1960).

For a non-interacting system at T = 0 the chemical potentials for the specis are given by: ($\hbar = c = 1$)

$$\mu_a = (p_{Fa}^2 + m_a^2)^{1/2} . \qquad (16.4)$$

According to (16.2) the thresholds for the emergence of the neutral baryons such as Λ, Σ^o, and Δ^o expressed in terms of the neutron number densities n_n are:

$$n_n(a) = (3\pi^2)^{-1}(m_a^2 - m^2)^{3/2} = \begin{cases} 0.96\ \text{fm}^{-3} & \text{for } \Lambda, \\ 1.75\ \text{fm}^{-3} & \text{for } \Sigma^o, \\ 2.23\ \text{fm}^{-3} & \text{for } \Delta^o. \end{cases} \qquad (16.5)$$

For positively charged baryons, we need to know the electron chemical potential, for which we may take the extreme relativistic approximation, $\mu_e \simeq p_{Fe}$. Charge neutrality of the system demands the equality of the proton Fermi momentum p_{Fp} to p_{Fe} (neglecting all other charged particles). Thus, the threshold for the emergence of a charged specis with mass m_a is determined by the conditions:

$$m_a = (p_{Fe}^2 + m^2)^{1/2} ,$$

and,

$$(p_{Fe}^2 + m^2)^{1/2} + p_{Fe} = (p_{Fn}^2 + m^2)^{1/2}, \qquad (16.6)$$

from which we find,

$$p_{Fn}^2(a) = 2\{(m_a^2 - m^2) + m_a(m_a^2 - m^2)^{1/2}\}. \qquad (16.7)$$

The threshold (neutron) densities are:

$$n_n(a) = (3\pi^2)^{-1} p_{Fn}^3(a) = \begin{cases} 20.6 \text{ fm}^{-3} & \text{for } \Sigma^+ , \\ 25.6 \text{ fm}^{-3} & \text{for } \Delta^+ . \end{cases} \qquad (16.8)$$

For negatively charged baryons, we obtain similarly the relations:

$$p_{Fn}^2(a) = 2\, p_{Fe}\{p_{Fe} + (p_{Fe}^2 + m^2)^{1/2}\},$$

where,

$$p_{Fe} = \frac{1}{3}\{2m_a - (m_a^2 + 3m^2)^{1/2}\} . \qquad (16.9)$$

The threshold (neutron) densities are:

$$n_n(a) = (3\pi^2)^{-1} p_{Fn}^3(a) = \begin{cases} 0.61 \text{ fm}^{-3} & \text{for } \Sigma^- , \\ 0.75 \text{ fm}^{-3} & \text{for } \Delta^- . \end{cases} \qquad (16.10)$$

For comparison, we note that the nucleon density of nuclear matter is $n_B = 0.19 \text{ fm}^{-3}$.

In order to study baryonic matter properly particle interactions must be included. Unfortunately, our knowledge of the hyperon interactions is far from complete making a realistic study of the problem difficult. In one of the earlier studies, Pandharipande (1971) applied the constrained variational method to a system of nucleons and hyperons. In this study, hyperon-hyperon and hyperon-nucleon interactions are assumed to be exactly the same as the nucleon-nucleon interactions, which are further assmed to be purely central potentials. This approach has been discussed in Section 13, and such assumptions are known to be rough. When Canuto (1975) applied the assumed potenitals for ΛN interaction to calculate ΛN scattering cross-section, he found poor agreement with experimental results.

Moszkowski (1974) introduced the following construction of the hyperon potentials. It is believed that the intermediate range of the NN potential is dominated by the exchange of quanta with the net quantum numbers of the σ-meson. We shall refer to the process simply as the exchange of the σ-meson. Assuming particle interactions to possess unitary symmetry, the σ-Λ and σ-Σ couplings would be in direct relations to the σ-N coupling. The situation can most easily be discussed in terms of the postulated quark contents of these particles. According to the quark model of particle substructures, the baryons are the bound states of three quarks, while the mesons are the bound states of quark-antiquark pairs. Baryons and mesons which are of interest to the present study are postulated to be composed of quarks of three flavors, denoted by u (up), d (down), and s (strange). The quark compositions of the following baryons are: $p = (u,u,d)$, $n = (u,d,d)$, $\Lambda = (u,d,s)$, $\Sigma^+ = (u,u,s)$, $\Sigma^o = (u,d,s)$, $\Sigma^- = (d,d,s)$. Each hyperon contains one strange quark. The isotriplet pions are given by: $\pi^+ = (u,\bar{d})$, $\pi^- = (d,\bar{u})$, and $\pi^o = (2)^{-1/2}(u\bar{u} - d\bar{d})$. The quark symbols are used in a way as if they are representing the quark wave functions forming the bound states. The σ-meson which is an isosinglet is given by: $\sigma = (2)^{-1/2}(u\bar{u} + d\bar{d})$. Describing the quarks by field operators and the antiquarks by the adjoint operators, the quark-antiquark wave function of the σ-meson acts essentially like a number operator for non-strange quarks. The coupling constants for the σ-baryon-baryon vertex are simply proportional to the numbers of non-strange quarks in the baryons. Since the nucleons contain three non-strange quarks while Λ and Σ hyperons contain two, the coupling constants $g_\sigma(\Lambda)$ and $g_\sigma(\Sigma)$ for the σ-Λ-Λ vertex and σ-Σ-Σ vertex, respectively, are only 2/3 of $g_\sigma(N)$ for the σ-N-N vertex. Consequently, if the intermediate range of the baryon-baryon interaction is dominated by the exchange of the σ-meson, the strength of the baryon-baryon potentials in this range should be related as follows:

$$V(\Lambda N) = V(\Sigma N) = \frac{2}{3} V(NN) ,$$

$$V(\Lambda\Lambda) = V(\Lambda\Sigma) = V(\Sigma\Sigma) = \frac{4}{9} V(NN) . \qquad (16.11)$$

Moszkowski therefore altered the Reid potentials in the above proportions and applied them to study baryonic matter containing hyperons. The many-body technique that he employed was the constrained variational method of Section 13. He found that Σ^- began to appear at densities as low as $n_B \approx 0.16 - 0.19 \text{ fm}^{-3}$, which is substantially lower than those estimated from free baryons given by (16.10).

The Δ coupling constants are again unknown, and they are not even

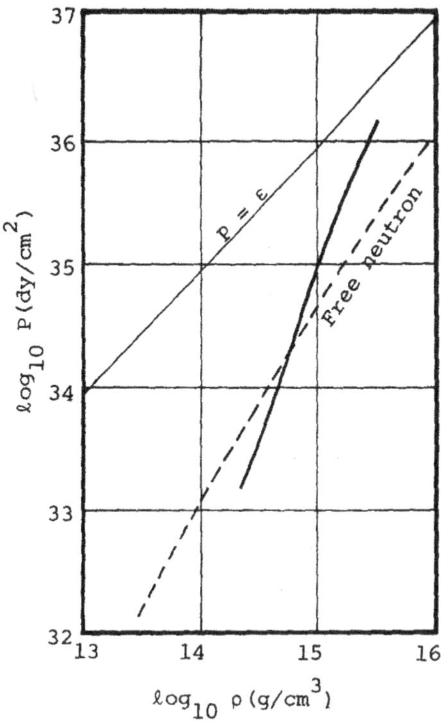

Figure 16-1. Equation of state of baryonic matter by Moszkowski (solid line).

in the same unitary symmetry multiplet with the nucleons. In Moszkowski's treatment, they are approximated by having the same coupling to the σ-meson as the nucleons, since the Δ's contain like the nucleons three non-strange quarks. The equation of state of baryonic matter thus computed is shown in Figure 16-1. Moszkowski also gave the following fits to the energy per baryon, E/A, and the pressure, P, in terms of the baryon number density n_B (quoted in Canuto, 1974):

$$E/A - 107.7\, n_B^{2/3} - 189.0\, n_B + 241.3\, n_B^{5/3} \,,$$

$$P = \{115.0\, n_B^{5/3} - 302.8\, n_B^{2} + 644.3\, n_B^{8/3}\} \times 10^{33} \,, \qquad (16.12)$$

where energy is in MeV, n_B in particles per fm^3, and pressure in dynes-/cm^2.

The Walecka model described in Section 14 may also be extended to study baryonic matter. In this model the two-nucleon interactions are assumed to be due to the exchange of (i) the σ-meson which generates the intermediate range attractive interaction, and (ii) the ω-meson which generates the short range repulsive interaction. In a way similar to Moszkowski's approach, the σ-meson coupling to the baryons may be assumed to be:

$$g_{\sigma\Lambda} = g_{\sigma\Sigma} = \frac{2}{3} g_{\sigma N} = \frac{2}{3} g_{\sigma\Delta} . \tag{16.13}$$

On the other hand, the ω-meson is assumed to couple universally to all baryons. (This amounts to saying in a potential approach that the hard core radii of all baryons are the same.) The ω-meson is assumed to couple to the baryon number of the particle, or its baryonic charge, which is the same for all baryons listed here. In this interpretation of the role of the ω-meson, it would produce a repulsive interaction between a pair of baryons, but attractive interaction between baryon and antibaryon. The mechanism generating this effect is the same as that of the photon, which induces repulsion between like charges but attraction between unlike charges.

Referring now to (14.66), we see that the chemical potentials of the baryons at T = 0 in the Walecka model are given by:

$$\mu_a = U_0^H + \sqrt{p_{Fa}^2 + m_{Ha}^2} , \tag{16.14}$$

where U_0^H is the repulsive interaction energy due to the exchange of the ω-meson and will be the same for all baryons, and m_{Ha} are the effective masses of the baryons. In analogy to (14.45), m_{Ha} are given by:

$$m_a - m_{Ha} = \sum_b (2\pi)^{-3} g_{\sigma a} g_{\sigma b} \gamma_b \int_0^{k_{Fb}} \frac{d^3p \; m_{Hb}}{\sqrt{\vec{p}^2 + m_{Hb}^2}} , \tag{16.15}$$

where the summation is over all specis of baryons composing the system. The hyperons having a smaller coupling constant will also have their effective masses closer to their actual masses than the nucleons. Since the effective mass is always lower than the actual mass, the matter density thresholds for the appearance of hyperons and Δ's are lower than those given in (16.5), (16.8) and (16.10).

The solution of the effective masses from (16.15) can be a rather tedious process since there are quite a few self-consistency conditions to be satisfied. They are due to the equality of the chemical potentials and charge neutrality of the system. In practice, however, fairly rapid convergence to the correct values can be achieved from an iteration procedure, since the effective masses are relatively stable parameters and it is possible to assume from the start that the mass shift $\Delta m = m - m_H$ for bayons to be about 2/3 of that for nucleons. The equation of state compted from the Walecka for baryonic matter with parameter given by (16.13) is shown in Figure 16-2.

At higher densities, more and more specis of baryons will emerge. The physical reason for the proliferation of particle specis is not known. Particle specis increases roughly like some power of the mass. In a form of particle dynamics called the Dual Resonance Model, an exponentially rising particle spectrum is related directly to particle interaction at high energies, and in a way it can be demonstrated that the inclusion of all the particles in the spectrum will account for partially particle interactions (Dashen, Ma and Bernstein 1969). Applications of this idea are however only valid for density domain which is much higher than that consider in this book, and therefore we shall not go into details of such an approach here. A summary of results from such an approach is given in Canuto (1975).

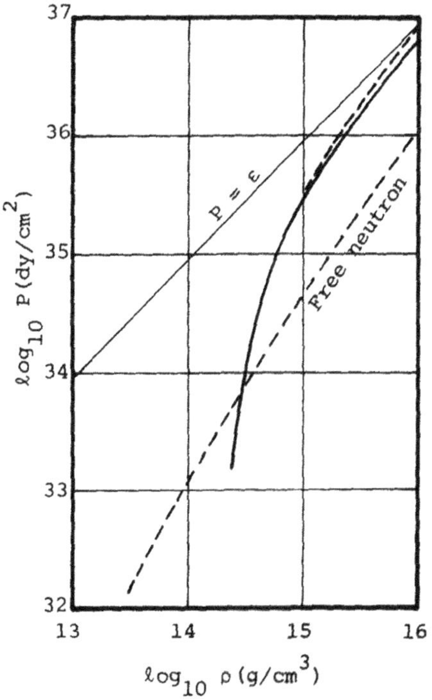

Figure 10-2. Equation of state of baryonic matter computed from the Walecka model (solid line); dotted extension is for pure neutron matter.

References

Ambartsumyan, V.A. and Saakyan, G.S. (1960). Astr. Zhu. 37, 193. (English Trans. Soviet Astr. 4, 187.)

Canuto, V. (1974). Ann Rev. Astron. Astrophys. 12, 167.

Canuto, V. (1975). Ann. Rev. Astron. Astrophys. 13, 355.

Dashen, R., Ma, S.K. and Bernstein, H.J. (1969). Phys. Rev. 187, 345.

Moszkowski,S.A. (1974). Phys. Rev. D9, 1613.

Pandharipande, V.R. (1971). Nucl. Phys. A174, 641.

17. Pion Condensation

As we have discussed before, a macroscopic body of matter at densities near the nuclear density will consist mainly of neutrons with very little admixture of protons, because protons must be accompanied by an equal number of electrons in order to preserve charge neutrality of the system,and since electrons are very energetic the ground state configuration of the system tends to exclude them and consequently protons. However, the situation will be different if protons were neutralized not by electrons but by negatively charged pions. Pions are much more massive than electrons and therefore their production threshold is high, but once they are produced pions being bosons will not contribute a Fermi energy to the system. They may all occur by a single momentum state as in a condensate. The presence of the pion condensate will modify the ground state structure of the neutron-proton system and its ground state energy.

The questions that concern us in this section are related to the issues whether or not it would be energetically favorable for such a pion condensate to appear, and if so what would be its threshold density and its effect on the equation of state. If pions were weakly interacting with the nucleons like the electrons, the transition from electrons to pions would occur as soon as the Fermi energy of the electrons equal to the rest mass of the pion, which is $m_\pi c^2 = 139.6$ MeV, but pions are hadrons and they interact strongly with the nucleons. The interaction of the pions with the background nucleons will modify their effective mass, and the pion condensate may appear at densities lower than that estimated from the free pion mass.

In low energy pion-nucleon scattering, the scattering amplitude may be decomposed into partial waves. S-wave pion-nucleon interaction is known to be repulsive. Hence, if the condensed pions interact with the background nucleon mainly via s-wave interaction, we would expect on very general grounds their effective mass would be higher than their rest mass, and we would find a higher threshold density for condensation to occur. Judging from the s-wave scattering lengths, the effective mass of a pion in a pure neutron matter may be raised by 38 MeV (see, for instance, Baym 1978). Our concern over the pion condensation problem is due to the strong attractive p-wave pion-nucleon interaction, which is known to lead to the (spin) $J = 3/2$ and (isospin) $T = 3/2$ resonance $\Delta(1236)$. We shall illustrate the pion condensation problem by concentrating on the pion-nucleon p-wave interaction.

For clarity of presentation we shall first ignore nucleon-nucleon interactions and consider just the interaction of the pions with the background nucleons. The dynamical problem can be expressed by the following Lagrangian:

$$L = -\int d^3x\{ \psi_n^+\left(-\frac{1}{2m}\nabla^2\right)\psi_n + \psi_p^+\left(-\frac{1}{2m}\nabla^2\right)\psi_p + \left(\partial_\mu\Phi^*\partial_\mu\Phi + m_\pi\Phi^*\Phi\right) + \\ + \sqrt{2}(f/m_\pi)\left(\psi_n\overleftrightarrow{\sigma}\cdot\vec{\nabla}\Phi\psi_p + \psi_p\overleftrightarrow{\sigma}\cdot\vec{\nabla}\Phi^+\psi_n\right)\} + L_e + L_\mu , \tag{17.1}$$

where ψ_n and ψ_p denote the neutron and proton fields, respectively, and are treated with non-relativistic kinematics, while Φ denotes the negative pion field and is treated relativistically. The interaction term is borrowed from the low energy pion-nucleon scattering theory. It gives rise to the p-wave interaction and predicts successfully the Δ resonance in an effective range approximation. f is called the rationalized coupling constant ($f^2/4\pi) \simeq 0.081$), and the factor $\sqrt{2}$ is carried over from isospin formalism which we do not choose to employ here. The electrons and muons are treated as free particles and their contributions to L, namely, L_e and L_μ, have not been written out explicitly.

Following the formalism described in Section 15, the equation of state of the pion field (from 15.31) is:

$$(-\partial_t^2 + \nabla^2 - m_\pi^2)\Phi = - f\,\vec{\nabla}\cdot\left(\psi_p\overleftrightarrow{\sigma}\psi_n\right)\sqrt{2}/m_\pi , \tag{17.2}$$

and its conjugate momentum (from 15.32) is:

$$\Pi = \partial_t\Phi^* . \tag{17.3}$$

The Hamiltonian (from 15.36) is:

$$H = \int d^3x\ \{\mathcal{H} + \mathcal{H}_e + \mathcal{H}_\mu\} ,$$

where,

$$H = \psi_n^+\left(-\frac{1}{2m}\nabla^2\right)\psi_n + \psi_p^+\left(-\frac{1}{2m}\nabla^2\right)\psi_p + \Pi^*\Pi + \Phi^*(-\nabla^2+m_\pi^2)\Phi +$$

$$+ \sqrt{2}(f/m_\pi)\left(\psi_n^+\overleftrightarrow{\sigma}\cdot\vec{\nabla}\Phi\psi_p + \psi_p^+\overleftrightarrow{\sigma}\cdot\vec{\nabla}\Phi^+\psi_n\right) , \qquad (17.4)$$

with H_e and H_μ given by free particle Hamiltonian densities.

Beta equilibrium under $n \to p + e + \nu$, and equilibrium under $n \leftrightarrow p + \pi^-$ imply the following relations among the chemical potentials of the particles:

$$\mu_n = \mu_p + \mu_e ,$$

$$\mu_n = \mu_p + \mu_\pi . \qquad (17.5)$$

Consequently,

$$\mu_\pi = \mu_e , \qquad (17.6)$$

which is the condition for the emergence of a pion condensate. Furthermore, if negative muons are present then we have $\mu_\mu = \mu_e$ as well.

The ground state energy E is the ground state expectation value of Hamiltonian H. It will be a function of the densities of the various particles present as well as the pion fields Φ and Π, which we shall treat as classical fields. A variation in the condensate density introduces the following change in the energy density:

$$\delta(E/\Omega) = \mu_\pi \, \delta n_\pi \qquad (17.7)$$

Expressing the variation in terms of the variations of the fields, we have:

$$\delta n_\pi = i\left(\Pi\delta\Phi^* + \Phi^*\delta\Pi\right) + \text{H.c.} ,$$

$$\delta(E/\Omega) = \delta\langle H\rangle = -\partial_t\Pi^*\delta\Phi + \partial_t\Phi^*\delta\Pi + \text{H.c.} , \qquad (17.8)$$

where H.c. denotes the Hermitian conjugate of the previous term. From these relations, the time dependence of the pion fields are shown to be:

$$\partial_t\Phi = -i\,\mu_\pi\Phi , \qquad \partial_t\Pi = -i\,\mu_\pi\Pi . \qquad (17.9)$$

Applying (17.3) and (17.9) we may replace the dependence of E on Π by μ_π and Φ, so that E is a function of the particle densities n_a as well as Φ and μ_π. These quantities are determined by minimizing the ground state energy with the constraint that the system be electrically neutral. Since particles can transmute into each other, it is more convenient to use the chemical potentials of the various particles as independent variables. This can be accomplished by means of the following Legendre transformation (Baym and Flower 1974):

$$\mathcal{H}' = \mathcal{H} - \mu_n n_n - \mu_p n_p - \mu_\pi n_\pi - \mu_e n_e - \mu_\mu n_\mu . \tag{17.10}$$

The ground state expectation value of $H' = \int d^3x\, \mathcal{H}'$ is denoted by E', and we have:

$$E'/\Omega = E/\Omega - \mu_n n_B + \mu_\pi n_q , \tag{17.11}$$

where Ω is the volume occupied by the system,

$$n_B = n_n + n_p ,$$

$$n_q = n_p - n_\pi - n_e - n_\mu . \tag{17.12}$$

Let the pion field be given by a simple trial wave function in the form of a running wave:

$$\Phi = \Phi_0\, e^{i\vec{k}\cdot\vec{x}} , \tag{17.13}$$

and in particular we choose $\vec{k}$ to be along the z-axis. The purpose of the exercise is to find the threshold value of $k = |\vec{k}|$ for which pion condensation occurs. Decomposing the nucleon fields in plane wave states as in (15.70), it is not difficult to see that the ground state expectation value of the Hamiltonian may be expressed as:

$$E'/\Omega = 2\sum_{p}(p^2/2m - \mu_n)u_n^+(\vec{p})u_n(\vec{p}) + 2\sum_{\vec{p}}(p^2/2m - \mu_p)u_p^+(\vec{p})u_p(\vec{p}) +$$

$$- 2\sum_{\vec{p}}\sqrt{2}\,\{iM_k\, u_n^+(\vec{p})\,\sigma_z\, u_p(\vec{p}-\vec{k})\,\Phi + \text{H.c.}\} + (-\mu_\pi^2+m_\pi^2+k^2)|\Phi|^2 , \tag{17.14}$$

where $M_k = fk/m_\pi$, and the last term is due to:

$$n_\pi = i(\Phi^*\Pi - \Pi^*\Phi) = 2\mu_\pi\,|\Phi|^2 . \tag{17.15}$$

(17.14) may be reexpressed in terms of uncoupled quasi-particle wave functions. Let,

$$U(\vec{p}) = \cos\theta\, u_n(\vec{p}) + i \sin\theta\, \sigma_z\, u_p(\vec{p}-\vec{k}) \;,$$

$$V(\vec{p}) = \cos\theta\, u_p(\vec{p}-\vec{k}) + i \sin\theta\, \sigma_z\, u_n(\vec{p}) \;, \qquad (17.16)$$

where θ is a parameter, then,

$$E'/\Omega = 2\sum_{\vec{p}}\left(u^+(\vec{p})U(\vec{p})E_p^- + V^+(\vec{p})V(\vec{p})E_p^+\right) + (-\mu_\pi^2 + m_\pi^2 + k^2)|\Phi|^2, \qquad (17.17)$$

where,

$$E_p^{\pm} = \frac{1}{2}\{\varepsilon_n - \varepsilon_p \pm \left(8M_k|\Phi|^2 + (\varepsilon_n-\varepsilon_p)^2\right)^{1/2}\} \;, \qquad (17.18)$$

with

$$\varepsilon_n = p^2/2m - \mu_n \;, \text{ and } \; \varepsilon_p = (\vec{p}-\vec{k})^2/2m - \mu_p, \qquad (17.19)$$

if θ is chosen to satisfy:

$$\tan 2\theta = \frac{2\sqrt{2}\,\Phi\, M_k}{\varepsilon_p - \varepsilon_n} \;. \qquad (17.20)$$

The ground state of the system is composed of all quasi-particle states for which E_p are non-positive (since the energy density of the system is already lowered by $\mu_n n_n$ and $\mu_p n_p$). The ground state energy is given by:

$$E'/\Omega = 2\sum_{\vec{p}}\{E_p^+\Theta(-E_p^+) + E_p^-\Theta(-E_p^-)\} + (-\mu_\pi^2+m_\pi^2+k^2)|\Phi|^2 \;, \qquad (17.21)$$

where the Θ-function is the step function given by (4.33).

Imposing the equation of motion of the pion field amounts to setting:

$$\frac{\partial E'}{\partial|\Phi|^2} \equiv F(\mu_n,\mu_\pi,k,\Phi) = 0 \;, \qquad (17.22)$$

which can be seen from differentiating (17.10). F is given by:

$$F(\mu_n,\mu_\pi,k,\Phi) = -\mu_\pi^2 + k^2 + m_\pi^2 + 4M_k^2\sum_{\vec{p}}\left(\frac{\Theta(-E_p^+) - \Theta(-E_p^-)}{E_p^+ - E_p^-}\right) . \qquad (17.23)$$

The threshold of condensate formation is evaluated from (17.22) in the limit $\Phi \to 0$. At this limit,

$$F(\mu_n,\mu_\pi,k,0) = (-\mu_\pi^2+k^2+m_\pi^2) + 4M_k^2 \sum_{\vec{p}} \frac{\Theta(\mu_n-p^2/2m) - \Theta(\mu_p-(\vec{p}-\vec{k})^2/2m)}{\frac{p^2}{2m} - \frac{(\vec{p}-\vec{k})^2}{2m} - \mu_\pi} .$$

Converting the summation over states into an integral over the momentum space, we find;

$$\sum_{\vec{p}} \frac{\Theta(\mu_n-p^2/2m)}{\frac{p^2}{2m} - \frac{(\vec{p}-\vec{k})^2}{2m} - \mu_\pi} = \frac{1}{(2\pi)^2} \int d\cos\theta \; p^2 dp \frac{2m}{2pk\cos\theta - k^2 - 2m\mu_\pi} =$$

$$= \frac{1}{(2\pi)^2} \frac{m}{k} \left\{ -\frac{1}{2} \left[p_{Fn}^2 - \left(\frac{k^2+2m\mu_\pi}{2k}\right)^2 \right] \ell og \frac{k^2+2m\mu_\pi+2kp_{Fn}}{k^2+2m\mu_\pi-2kp_{Fn}} - \frac{k^2+2m\mu_\pi}{2k} p_{Fn} \right\}$$

(17.24)

where $p_{Fn}^2 = 2m\mu_n$. Similar result is obtained for the other summation. Setting $F(\mu_n,\mu_\pi,k,0) = 0$, we obtainthe threshold condition:

$$\mu_\pi^2 = k^2+m_\pi^2 - \frac{M_k m}{2k^2\pi^2} \left(\frac{4k^2p_{Fn}^2-(k^2+2m\mu_\pi)^2}{4k} \ell og \left| \frac{2m\mu_\pi+k^2+2kp_{Fn}}{2m\mu_\pi+k^2-2kp_{Fn}} \right| + (k^2+2m\mu_\pi)p_{Fn} + \right.$$

$$\left. + \frac{4k^2p_{Fp}^2-(k^2-2m\mu_\pi)^2}{4k} \ell og \left| \frac{-2m\mu_\pi+k^2+2kp_{Fp}}{-2m\mu_\pi+k^2-2kp_{Fp}} \right| + p_{Fp}(k^2-2m\mu_\pi) \right) \quad (17.25)$$

This is the condition derived by Baym and Flower (1974). Since $p_{Fp}^2 = 2m(\mu_n-\mu_\pi)$, (17.25) is in fact a relation between μ_n, μ_π and k. We shall introduce p_F which is related to the total nucleon number density by $p_F = (3\pi^2 n_B)^{1/3}$, or $p_F^3 = p_{Fn}^3 + p_{Fp}^3$, and plot the pion threshold region in the k - p_F plane. This is shown in Figure 17-1. There is no pion condensation to the left of the curve (F > 0). Pion condensate can form to the right of the curve. With increasing p_F the first point where pions condense correspond to the point on the curve with a vertical tangent to the left. The condensate has a finite wave vector k. The values of μ_n, k and n_B at the earliest point of condensation are estimated from this study to be:

$$\mu_\pi^c = 29.8 \text{ MeV} , \; k^c = 1.1 \text{ fm}^{-1} , \; n_B^c = 0.058 \text{ fm}^{-3}. \quad (17.26)$$

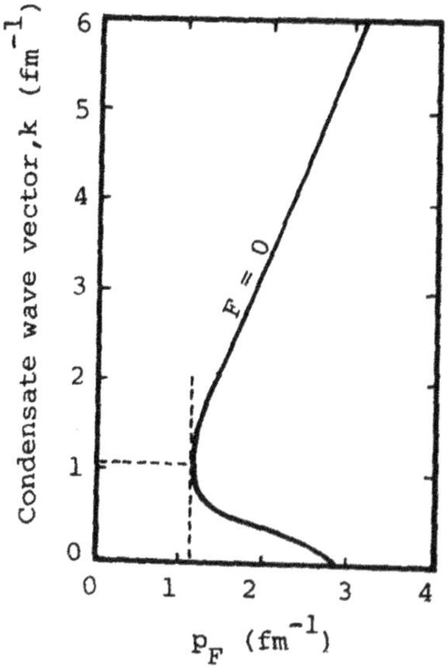

Figure 17-1. Pion condensation threshold curve in the k vs. p_F plane.

It shows that pion condensation occurs at a density which is about one-third of the nuclear matter density. The estimated threshold density is alarmingly low, and much theoretical efforts had subsequently been directed to verify this result. A variety of interactions had been applied by Au and Baym (1974) to reinvestigate this problem (see also Au 1976). A simplified form of nucleon interaction which depends only on the total baryon density is included. Typical results are as follows: $n_B^c = 0.856\ \text{fm}^{-3}$, $k^c = 2.53\ \text{km}^{-1}$, and $\mu_\pi^c = 377$ MeV, which are considerably higher than those listed in (17.26).

In general the effect of condensation on the equation of state is very sensitive to the structure and magnitude of the p-wave pion nucleon interaction. A 10% reduction in the coupling constant f^2 could raise the condensation threshold density by a factor of two, and this is illustrated by comparing model VIII and model IX of Au and Baym (1974). A typical equation of state with pion condensation is shown in Figure 17-2, which is due to Weise and Brown (1975). The flat portion in the equation of state indicates a first-order phase transition in the system.

We shall next look at a model of pion condensation which includes nucleon-nucleon interactions and treat them on an equal footing. This is achieved by extending the mean field model of Section 15. The Lagran-

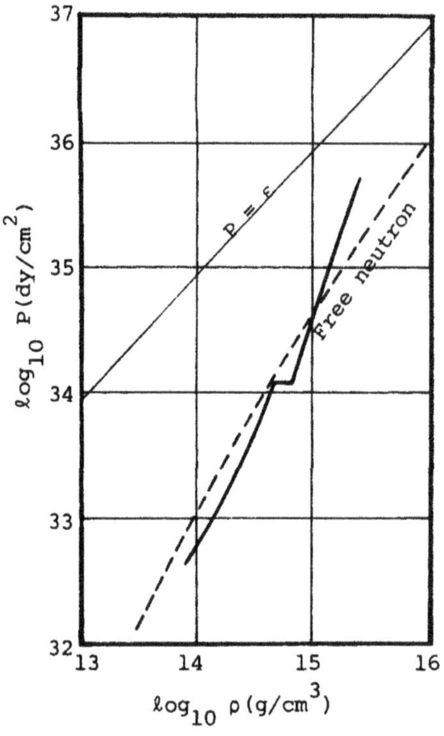

Figure 17-2. Equation of state of pion condensation due to Weise and Brown (1975).

gian density is enlarged by adding the pion part:

$$L = L_N^o + L_\omega^o + L_\sigma^o + L_\omega^I + L_\sigma^I + L_\pi , \tag{17.27}$$

where the first five terms on the right are given by (15.38), while,

$$L_\pi = -\frac{1}{2}\{ \partial_\mu\vec{\Phi}\partial_\mu\vec{\Phi} + m_\pi^2 \vec{\Phi}^2\} + ig_\pi\left(\hat{\psi}^+\gamma_4\gamma_5\gamma_\mu\vec{\tau}\hat{\psi}\right) \partial_\mu\vec{\Phi}. \tag{17.28}$$

$\vec{\Phi}$ denotes the pion field and the vector symbol is to indicate that it is an isovector. This Lagrangian density differs from that of (15.37) in that the interaction is no longer diagonal in the isospin space, and $\vec{\tau}$ will couple the two isospin states of the nucleon. The pion-nucleon interaction is taken to be pseudovector $(\gamma_5\gamma_\mu)$, which gives rise to the expected p-wave interaction at low energy.

We shall again adopt the mean field approach and replace the meson fields by their expectation values in the nucleon ground state. The vector components of V_μ will again vanish leaving $V_0 \neq 0$, which together with ϕ are given by constants independent of space and time. The pion mean field will be given a spatial dependence, which is written in the form of a running wave;

$$\vec{\Phi} = \Phi_0 (\hat{u} \cos k\cdot x + \hat{v}\times\hat{u} \sin k\cdot x) , \tag{17.29}$$

where $\hat{u}$ and $\hat{v}$ are othronomal vectors in the isospin space, and:

$$k\cdot x = \vec{k}\cdot\vec{x} - k_0 t , \tag{17.30}$$

and as it has been demonstrated before $k_0 = \mu_\pi$. The pion mean field is given a more general isospin expression than it was before. It returns to the form (17.14) if $\hat{u}$ is along the x-direction and $\hat{v}$ in the z-direction of the isospin space. This means:

$$\vec{\tau}\cdot\vec{\Phi} = \tau_+ e^{ik\cdot x} + \tau_- e^{-ik\cdot x} , \tag{17.31}$$

where $\tau_\pm = \frac{1}{2}(\tau_x \pm i\tau_y)$. The equation of motion for the pion mean field is:

$$(\partial_\mu\partial_\mu - m_\pi^2)\vec{\Phi} = -ig_\pi \partial_\mu \vec{J}_\mu ,$$

where,

$$\vec{J}_\mu = \langle\Psi_0| \hat{\psi}^+ \gamma_4\gamma_5\gamma_\mu \vec{\tau} \hat{\psi} |\Psi_0\rangle . \tag{17.32}$$

The equation of motion for the nucleon field is:

$$\{ \gamma_\mu\partial_\mu + g_\omega\gamma_4 V_0 + m - g_\sigma\phi - ig_\pi\gamma_5\gamma_\mu\partial_\mu(\vec{\tau}\cdot\vec{\Phi}) \} \hat{\psi}(x) = 0 , \tag{17.33}$$

which depends explicitly on space-time through the pion field. However, this dependence can be transformed into a phase factor. We find that we can rewrite,

$$\vec{\tau}\cdot\vec{\Phi} = \Phi_0 R \vec{\tau}\cdot\hat{u} R^+, \tag{17.34}$$

where R is a space-time dependent rotation in the isospin space,

$$R = \exp\{ -\frac{1}{2} i(k\cdot x) \vec{\tau}\cdot\hat{v} \} . \tag{17.35}$$

Also, the pion term in the Dirac equation is replaced by:

$$\partial_\mu(\vec{\tau}\cdot\vec{\Phi}) = k_\mu\Phi_0 R(\vec{\tau}\cdot\hat{v}\times\hat{u})R^+ . \tag{17.36}$$

Thus, by defining a rotated nucleon field by:

$$\psi' = R^+ \psi , \tag{17.37}$$

which we shall refer to as a quasi-particle, it satisfies a Dirac equation which has simple plane wave solutions:

$$\{ \gamma_\mu\partial_\mu + g_\omega\gamma_4V_0 + m - g_\sigma\phi - i(\gamma_\mu k_\mu)\vec{\tau}\cdot(\tfrac{1}{2}\hat{v} + g_\pi\gamma_5\hat{v}\times\hat{u}\Phi_0)\}\hat{\psi}' = 0. \quad (17.38)$$

Writing $\hat{\psi}'$ as:

$$\hat{\psi}' = w(\vec{p})a_{\vec{p}}e^{ip\cdot x} , \quad (17.39)$$

where $w(\vec{p})$ is an eight-component vector in both the Dirac spinor and isospin spaces, and $a_{\vec{p}}$ is an annihilator operator for a quasi-particle. $w(\vec{p})$ satisfies:

$$\{i\gamma_\mu P_\mu - m^* - i\gamma_\mu k_\mu\vec{\tau}\cdot(\tfrac{1}{2}\hat{v} + g_\pi\Phi_0\gamma_5\hat{v}\times\hat{u})\} w(\vec{p}) = 0 , \quad (17.40)$$

where,

$$P_\mu = p_\mu - g_\omega V_0(i\delta_{\mu 4}) .$$

and,

$$m^* = m - g_\sigma\phi . \quad (17.41)$$

The eigenvalue p_0 of (17.40) can be found by multiplying the Dirac equation by the inverse of the operator acting on $w(\vec{p})$. It is given by the propagator (Banerjee, Glendening, Gyulassy 1981):

$$S(p) = \{i\gamma_\mu P_\mu - m^* + \gamma_\mu k_\mu\vec{\tau}\cdot(\tfrac{1}{2}\hat{v} + g_\pi\Phi_0\gamma_5\hat{v}\times\hat{u})\}^{-1}$$

$$= \frac{1}{D(p)}\{ -P\cdot P - \varepsilon^2 - (P\cdot k)\vec{\tau}\cdot\hat{v} + i2g_\pi\Phi_0\big(i(P\cdot k) + m^*\gamma_\mu k_\mu\big)\gamma_5\vec{\tau}\cdot\hat{v}\times\hat{u}\} \times$$

$$\times \{i\gamma_\mu P_\mu + m^* - i\gamma_\mu k_\mu\vec{\tau}\cdot(\tfrac{1}{2}\hat{v} - g_\pi\Phi_0\gamma_5\hat{v}\times\hat{u})\} , \quad (17.42)$$

where,

$$D(p) \equiv D(p_0,\vec{p}) = (P\cdot P + \varepsilon^2)^2 - (P\cdot k)^2 - 4g_\pi^2\Phi_0^2\big((P\cdot k)^2 + m^{*2}(k\cdot k)\big) , \quad (17.43)$$

and,

$$\varepsilon^2 = m^{*2} + (\tfrac{1}{4} + g_\pi^2\Phi_0^2)(k\cdot k). \quad (17.44)$$

Applying $S(p)$ onto (17.40) would make $w(\vec{p}) = 0$ unless $D(p)$ vanishes. Consequently, the quasi-particle spectrum $p_0 = \omega(\vec{p})$ is given by the solution of:

$$D(\omega(\vec{p}),\vec{p}) = 0 , \quad (17.45)$$

which is a fourth order equation in p_0. The case related to symmetric nuclear matter with $n_n = n_p$ has been studied by Banerjee et al. (1981). In this case, the solution of (12.45) for p_0 is particularly simple, since

$$k_0 = \mu_n - \mu_p = 0 \,. \tag{17.46}$$

The positive solutions of P_0 are found to be:

$$P_0 = E_\pm = \sqrt{\vec{P}^2 + \varepsilon^2 \pm \Delta^2} \,, \tag{17.47}$$

where,

$$\Delta^2 = \{ (1+4g_\pi^2\Phi_0^2)(\vec{P}\cdot\vec{k})^2 + (2g_\pi\Phi_0 m_\pi^*)^2\vec{k}^2 \}^{1/2} \tag{17.48}$$

Hence, the quasi-particle spectrum is:

$$\omega(\vec{p}) = g_\omega V_o + E_\pm \,. \tag{17.49}$$

The energy density of the system is given by:

$$E/\Omega = 2(2\pi)^{-3}\int d^3p \, \{(E_- + g_\omega V_o)\Theta_- + (E_+ + g_\omega V_0)\Theta_+\} + + \frac{1}{2}(\vec{k}^2 + m_\pi^2)\Phi_0^2 - \frac{1}{2}m_\omega^2 V_0^2 + \frac{1}{2}m_\sigma^2\phi^2 + U(\phi) \,, \tag{17.50}$$

where,

$$\Theta_\pm = \Theta(E_F - E_\pm) \,, \tag{17.51}$$

and $U(\phi)$ denotes the non-linear terms (15.78). The nucleon number density is given by:

$$n_B = 2(2\pi)^{-3}\int d^3p \, \{\Theta_- + \Theta_+\} \,. \tag{17.52}$$

The source term $\vec{J}_\mu$ for the pion field equation of motion can similarly be evaluated, and so can the source term for the σ-meson field. These we shall not write out here. The idea of this approach is to seek self-consistent solutions of Φ_0 and ϕ from these equations and use them to evaluate E/Ω from (17.50). From our previous analysis of the Walecka model, it is known that $\Phi_0 = 0$ is a solution, and has been taken to be the solution at nuclear matter density. Banerjee et al. investigated the conditions for having a $\Phi_0 \neq 0$ solution at the same density, or in other words, the values of the meson parameters which will give rise to an abnormal state with a pion condensate that lowers the ground state energy. They found that for g_π sufficiently large, such an abnormal state did exist.

Applying the same procedure to investigate the effect of a pion condensate in dense matter would start with a fixed set of meson parameters. The proton to neutron ratio would not be restricted to one and hence $k_0 \neq 0$. The threshold density for the appearance of a pion condensate is determined by finding the lowest p_F for which a $\Phi_0 \neq 0$ solution that lowers the ground state energy appears. An equation of state for systems obeying the dynamics prescribed by the Lagrangian density (17.27) can thus be determined. Due to the uncertainty with some the meson parameters and the inherent difficulty in working with a strong interaction theory, the results of pion condensation can not be firmly established unless some independent experimental verifications are performed. We end this section with this note of caution.

References

Au, C.K. (1976). Phys. Lett. 61B, 300.

Au, C.K. and Baym, G. (1974). Nucl. Phys. A236, 500.

Banerjee, B., Glendenning, N.K. and Gyulassy, M. (1981). Nucl. Phys. A361, 326.

Baym, G. (1978). In Nuclear Physics with Heavy Ions and Mesons, (Balian, R., Rho, M. and Ripka, G., Eds.), Vol. 2, p.745, North-Holland, Amsterdam and New York.

Baym, G. and Flowers, E. (1974). Nucl. Phys. A222, 29.

Weise, W. and Brown, G.E. (1975). Phys. Lett. 58B, 300.

Bibliography

Baym, G. and Campbell, D.K. (1979). In Mesons in Nuclei, III (Rho, M. and Wilkinson, D. Eds), p.1031, North-Holland, Amsterdam and New York.

Brown, G.E. and Weise, W. (1976). Phys. Rep. 27C, 1.

Migdal, A.B. (1971). Ah ETF 61, 2210; JETP (Sov. Phys.) 34, 1184 (1972

Migdal, A.B. (1978). Rev. Mod. Phys. 50, 107.

Sawyer, R.F. (1972). Phys. Rev. Lett. 29, 382.

Scalapino, D.J. (1972). Phys. Rev. Lett. 29, 386.

18. Quark Matter

We infer from the charge form factor of the nucleon that it is an extended object. Early explanations of the form factor were in terms of the meson theory, which treats the nucleon as a point object and associates the form factor to the mesonic cloud surrounding it. Such theory achieved some degrees of success in the past. However, the inherent difficulty in working with strong interaction theory of this type limits its application. In recent years, with the advent of the quark model more and more attention has been directed towards a model of the nucleon as a bound state of quarks, even though no free quark has actually been identified. A quark possesses various attributes in addition to its mass and spin, such as the c-charge (color), f-charge (flavor) and electric charge. The dynamical description of the quarks in the strong interaction domain is based on quantum chromodynamics (QCD), which is a non-Abelian gauge theory, fully renormalizable and possessing interesting asymptotic limits.

In the quark model, the nucleon is viewed as the bound state of three quarks, which interact via the exchange of gluons as specified by QCD. The size of the nucleon is therefore determined by the confining radius of the quark interactions. When matter density reaches such a degree that the average separation per nucleon is less than the confining radius, the identity of the nucleon will be lost and matter at such densities should more appropriately be described as quark matter. Estimating the nucleon size to be given by a radius between 0.5 - 1.0 fm, which gives an indication of the confining radius, matter is expected to make transitions to quark matter at densities between 2 - 10 times the nuclear density.

The quarks are spin-½, fractionally charged fermions. They are usually labelled according to their flavors, such as u (up), d (down), s (strange), c (charm), and possibly others, such as b (bottom) and t (top). The electric charges of the first four types in units of the proton charge is 2/3, -1/3, -1/3, and 2/3, respectively. They each have baryon number 1/3 and all have strangeness zero except s which has strangeness -1. For each quark there is a corresponding anti-quark with conjugate quantum numbers.

Quarks interact weakly via the f-charges. Transitions among quarks of different flavors are made under the weak interactions;

$$d \to u + \ell + \bar{\nu} , \qquad u + \ell \to d + \nu ,$$

$$s \to u + \ell + \bar{\nu} , \qquad u + \ell \to s + \nu , \tag{18.1}$$

where ℓ denotes an electron or muon. Chemical equilibrium in a quark matter requires the following conditions among the chemical potentials of the specis;

$$\mu_d = \mu_u + \mu_e = \mu_s ,$$

$$\mu_e = \mu_\mu . \tag{18.2}$$

Furthermore, electrical neutrality of the system requires:

$$\frac{2}{3} n_u = \frac{1}{3} (n_d + n_s) + n_e + n_\mu . \tag{18.3}$$

Quarks interact strongly via their c-charges. In analogy to the isospin symmetry, which is described by the symmetry group SU(2), the c-charge interactions obey a symmetry described by the group SU(3), which is generated by a Lie algebra consisting of 8 elements, λ^α (α = 1, 2, ..., 8). The SU(3) generators are commonly assigned the following matrix forms:

$$\lambda^1 = \begin{pmatrix} 0 & 1 & 0 \\ 1 & 0 & 0 \\ 0 & 0 & 0 \end{pmatrix} , \qquad \lambda^2 = \begin{pmatrix} 0 & -i & 0 \\ i & 0 & 0 \\ 0 & 0 & 0 \end{pmatrix} , \qquad \lambda^3 = \begin{pmatrix} 1 & 0 & 0 \\ 0 & -1 & 0 \\ 0 & 0 & 0 \end{pmatrix} ,$$

$$\lambda^4 = \begin{pmatrix} 0 & 0 & 1 \\ 0 & 0 & 0 \\ 1 & 0 & 0 \end{pmatrix} , \qquad \lambda^5 = \begin{pmatrix} 0 & 0 & -i \\ 0 & 0 & 0 \\ i & 0 & 0 \end{pmatrix} , \qquad \lambda^6 = \begin{pmatrix} 0 & 0 & 0 \\ 0 & 0 & 1 \\ 0 & 1 & 0 \end{pmatrix} ,$$

$$\lambda^7 = \begin{pmatrix} 0 & 0 & 0 \\ 0 & 0 & -i \\ 0 & i & 0 \end{pmatrix} , \qquad \lambda^8 = \frac{1}{\sqrt{3}} \begin{pmatrix} 1 & 0 & 0 \\ 0 & 1 & 0 \\ 0 & 0 & -2 \end{pmatrix} . \tag{18.4}$$

The three c-charge states of the quark are represented by column matrices:

$$\begin{pmatrix}1\\0\\0\end{pmatrix}, \qquad \begin{pmatrix}0\\1\\0\end{pmatrix}, \qquad \begin{pmatrix}0\\0\\1\end{pmatrix}. \tag{18.5}$$

The λ matrices are Hermitian and traceless:

$$\lambda^{\alpha+} = \lambda^{\alpha} , \qquad \mathrm{Tr}(\lambda^{\alpha}) = 0 , \tag{18.6}$$

and are normalized by:

$$\mathrm{Tr}(\lambda^{\alpha}\lambda^{\beta}) = 2\delta_{\alpha\beta}. \tag{18.7}$$

The structure constants $f_{\alpha\beta\gamma}$ of the Lie algrbra are defined by:

$$[\lambda^{\alpha}, \lambda^{\beta}] = 2if_{\alpha\beta\gamma}\lambda^{\gamma} , \tag{18.8}$$

where $f_{\alpha\beta\gamma}$ are totally antisymmetric in their indices with:

$$\begin{aligned} &f_{123} = 1 , \\ &f_{147} = f_{246} = f_{257} = f_{345} = f_{516} = f_{637} = \frac{1}{2} , \\ &f_{458} = f_{678} = (3/4)^{1/2} . \end{aligned} \tag{18.9}$$

The others are either zero (with two or more equal indices) or negative of these (with indices permuted). The λ's also satisfy anticommutation relations, which are given by:

$$\{\lambda^{\alpha}, \lambda^{\beta}\} = \frac{4}{3}\delta_{\alpha\beta} + 2d_{\alpha\beta\gamma}\lambda^{\gamma} , \tag{18.10}$$

where $d_{\alpha\beta\gamma}$ are totally symmetric in their indices. The relevant components of d's are;

$$\begin{aligned} &d_{118} = d_{228} = d_{338} = -d_{888} = (1/3)^{1/2} , \\ &d_{146} = d_{157} = d_{256} = d_{344} = d_{355} = \frac{1}{2} , \\ &d_{247} = d_{366} = d_{377} = -\frac{1}{2} , \\ &d_{448} = d_{558} = d_{668} = d_{778} = -(1/12)^{1/2} . \end{aligned} \tag{18.11}$$

The others can be deduced from symmetry.

Denoting the quark fields as q_{ai} where a is the index for c-charge and i for f-charge, and the gluon fields by A_μ^α which is a vector field with vector index μ and c-charge index α, the Lagrangian density for the quark-gluon dynamics is given by;

$$L = \bar{q}_{ai}\gamma_\mu(i\delta_{ab}\partial_\mu + \frac{1}{2}g\lambda^\gamma_{ab}A^\gamma_\mu)q_{bi} - \bar{q}_{ai}\, m_{ij}\, q_{aj} + \frac{1}{2}F^\gamma_{\mu\nu}F^\gamma_{\mu\nu} ,$$

where

$$F^\gamma_{\mu\nu} = \partial_\mu A^\gamma_\nu - \partial_\nu A^\gamma_\mu + gf_{\gamma\alpha\beta}\, A^\alpha_\mu A^\beta_\nu \quad , \tag{18.12}$$

where $\bar{q} = q^+\gamma_4$. The Lagrangian density describes a theory which is invariant under the following gauge transformations of the quark fields:

$$q_{ai} \rightarrow (\delta_{ab} + i\lambda^\gamma_{ab}\,\theta^\gamma)\, q_{bi} , \tag{18.13}$$

where θ^γ are infinitesimal real parameters. The theory is renormalizable and because the generators of the gauge transformations, i.e. λ^γ, do not commute with each other, the theory called a non-Abelian gauge theory has interesting asymptotic limits. To understand the phenomemon let us first review the situation in quantum electrodynamics (QED), which is an Abelian gauge theory since there is a single vector field (the photon) mediating the interaction. The theory permits particle-antiparticle pairs to be spontaniously created and annihilated, which is referred to as vacuum polarization. A charged particle will be screened by the vacuum polarization process so that its observed charge is less than the coupling constant assigned by the theory. The observed charge is called the renormalized charge. In an actual computation of the renormalized charge, a sequence of particle-antiparticle loops inserted into the photon line is summed. It is a consequence of Fermi statistics that fermion-antifermion loops contributes a neagative sign to the process leading to a renormalized charge which is less than that given by the coupling constant. In other words, if the charge were probed at short distance penetrating the screening polarization we would detect a much larger coupling constant than the charge that we observe. Let us now turn to QCD, which being a non-Abelian theory possesses not one but eight gluon fields. Therefore in computing the renormalized c-charge, not only quark-antiquark loops but also gluon loops contribute to the process. Gluons being bosons their loops have the opposite sign to the quark-anti-quark loops. The gluon loops will give rise to "antiscreening" to the c-charge. c-charge is actually antiscreened if the number of quark flavors do not exceed 17, which is believed to be the case. Antiscreening will make the renormalized c-charge larger than the coupling constant. Hence if the coupling constant is probed at short distance, it is actually smaller. It is the nature of strong interaction that renormalization alters the properties not just perturbatively but categorically. Thus, at very short distance we may detect a coupling constant which is vanishingly small. This phenomenon is called asymptotic freedom. On the other hand, if quarks were to separate from each other they would have to overcome an ever-increasing renormalized coupling constant. Nucleons and mesons must therefore be color-singlets for they do not interact directly via c-charges but only indirectly as quark-antiquark pairs are produced.

In electrostatics, when a positive electric charge is placed inside a hole surrounded by a dielectric medium, it induces negative charge on the surface of the hole. The reason for this is the same as the screening effect. In analogy we can think of the antiscreening of the c-charge as a case where a charge induces charge of the same sign on the surface of the hole. The induced charge would therefore repel the initial charge, or in other words, a pressure is exerted by the surface of the hole on its content. In the case of the c-charge the region surrounding it can be delineated into two parts, the part close to the c-charge has a relatively small coupling constant (region of asymptotic freedom), while the part beyond that has a larger coupling constant and antiscreening is fully developed. Due to the non-linear nature of gluon coupling, we would expect the transition from one region to the other rather abrupt, making the analogy with a charge in the hole quite appropriate. For the c-charge the medium surround the hole is the polarized vacuum which behaves like a dia-electric medium (in the terminology of electrostatics) rather than a dielectric medium (Lee 1981).

Quark models of the nucleon which make use of this observation are called bag models. A bag is just the region containing the c-charges surrounded by the polarized vacuum. The reaction of the induced charges on the surface of the bag on its content is attributed to either (i) a volumetric energy density as in the MIT bag model, or (ii) a surface energy density, as in the SLAC bag model. In one case the bag energy is proportional to the volume of the bag while in the other to the surface area. The MIT bag model has been applied to study dense quark matter (Baym and Chin, 1976), and we shall discuss this case here.

In addition to the volumetric energy density B, the MIT bag model includes some residual quark interactions inside the bag (Chodo, et al. 1974). The residual interactions would be weak and may be treated as effective interactions as in the case of nucleons inside nuclear matter. They should not be neglected since baryon spectroscopy can be explained by just such residual interactions. Quarks inside a bag interact via the following effective Lagrangian density:

$$L = \bar{q}_{ai}(\gamma_\mu \partial_\mu + m_i) q_{ai} + \frac{i}{2} g\, \bar{q}_{ai}\, \lambda^{\alpha}_{ab}\, A^{\alpha}_{\mu}\, \gamma_\mu q_{bi}\ , \qquad (18.14)$$

In applying the bag model to quark matter, we assume that bags merge to form a giant bag of macroscopic size with no trace of the polarized vacuum left within the macroscopic body.

Even though free quarks are believed to be rather massive, since their production thresholds are above all experimental energies, the quark masses inside a bag may be rather low. Theyare estimated to be under 100 MeV for the u and d quarks, $\sim$100-200 MeV for the s quarks and $\sim$2 GeV for the c quarks. As long as quark Fermi energy in quark matter is high compared to the quark mass,it may be neglected. We shall eventually take this limit, but for the time being we shall retain the quark mass whenever the Dirac wave functions are employed. (18.14) neglects on the outset the gluon mass.

Quark matter will have zero net c-charges, and thus for every quark flavor, all three color states will be participating equally. The energy density (energy per unit volume) for quark matter in the MIT bag model is derived by Baym and Chin (1976) to be:

$$E/\Omega = B + \frac{3}{4\pi^2} \sum_i \left(1 + \frac{2\alpha_c}{3\pi}\right) p_{Fi}^4 , \quad (18.15)$$

where the summation is over all quark flavors and $\alpha_c = g^2/4\pi$. It consists of the bag volumetric energy B, the quark kinetic energy (first terms in summation) and interaction energy (second terms in summation). The kinetic energy is derived with the quark mass set equal to zero, and multiplied by a factor of three to account for the three color states. The interaction energy terms will be discussed below.

The interaction energy is computed from the effective Lagrangian density (18.14) in a manner similar to that described in Sections 14 and 15. The equations of motion for the quark fields give rise to the following Hartree-Fock equations: (see 14.30)

$$\{ \vec{\alpha}\cdot\vec{p} + \beta(m + U^H + U^F) \} u(\vec{p}) = \varepsilon_p^{HF} u(\vec{p}) , \quad (18.16)$$

where U^H and U^F arise out of the exchange of the gluons. In Section 14 U^H and U^F are given for interactions mediated by both the scalar σ-meson and the vector ω-meson. These results may be taken over readily if we drop the σ-meson terms and replace the ω-meson couplings by the gluon couplings. The repacements are:

$$g_\omega \rightarrow \frac{g}{2} (\lambda^\gamma)_{ab} , \quad (18.17)$$

where γ corresponds to the gluon charge while a(b) to the c-charge of the outgoing (incoming) quark. The direct interaction (or Hartree) terms would have a = b. If we examine the λ-matrices given in (18.4) we see that they have no diagonal elements, except for λ^3 and λ^8, and so do not contribute to the direct interaction processes. As to λ^3 and λ^8, they are traceless, so that when all three quark color states participate equally the final interaction is zero. Hence, $U^H = 0$ in (18.16). U^F is given by (14.32), which specializes to the present case as:

$$\beta U^F(p) = U_0^F(p) + \vec{\alpha}\cdot\hat{p}\, U_v^F(p) , \quad (18.18)$$

with (see 14.63):

$$U_0^F(p) = (2\pi)^{-3}\int d^3q\ G(\lambda) ,$$

and

$$U_V^F(p) = -(2\pi)^{-3}\int d^3q \frac{\hat{p}\cdot\vec{q}_V}{\sqrt{q_V^2+m_{HF}^2}} G(\lambda) , \qquad (18.19)$$

where, in the limit of zero gluon mass,

$$G(\lambda) = \sum_{\gamma,\gamma'=1}^{8} \left(\frac{g}{2}\right)^2 \mathrm{Tr}(\lambda^\gamma\lambda^{\gamma'}) \frac{1}{(p-q)_\mu^2} = \frac{4g^2}{(p-q)_\mu^2} . \qquad (18.20)$$

In (18.20), the summations over the color states of the quarks are replaced by the trace (Tr) of the λ's. Also, if the quark mass were set equal to zero, m_{HF} according to (14.65) would again be related to m_{HF} and is therefore proportional to higher orders in the coupling constants. Consequently, m_{HF} may also be set equal to zero. The interaction energy to order g^2 for each flavor of quark in the limit of zero quark mass is:

$$E_I/\Omega = \frac{1}{2}\times\frac{(4g^2)}{(2\pi)^6}\times 2\times\int d^3p d^3q \frac{u^+(\vec{p})\ (1-\vec{\alpha}\cdot\hat{q})\ u(\vec{p})}{(p-q)_\mu^2}$$

$$= \frac{g^2}{\pi^4}\int_0^{p_F} p^2dpq^2dq \frac{(1-\hat{p}\cdot\hat{q})}{2pq(1-\hat{p}\cdot\hat{q})} = \left(\frac{g^2}{4\pi}\right)\frac{1}{2\pi^3}p_F^4 , \qquad (18.21)$$

since $\hat{q} = \vec{q}_V/q_V$, $p\cdot p = q\cdot q = 0$ and $p_0 = |\vec{p}| = p$, $q_0 = |\vec{q}| = q$. This is the result of (18.15), which was originally derived from perturbation theory and therefore factors like m_{HF} and q_V do not appear. These differences disappear in the limit of zero quark mass.

The pressure of the quark matter is given by:

$$P = -B + (4\pi^2)^{-1}\sum_i \left(1+\frac{2\alpha_c}{3\pi}\right)p_{Fi}^4 . \qquad (18.22)$$

Baym and Chin (1976) also estimated the density at which a phase transition from neutron matter to quark matter occurs. This can be done by the tangent construction method described in Section 6. The energies per neutron at fixed volume per nucleon in the two phases are compared. This is shwon in Figure 18-1. Neutron energies in the nucleonic phase are taken from the calculations of Pandharipande (1971)

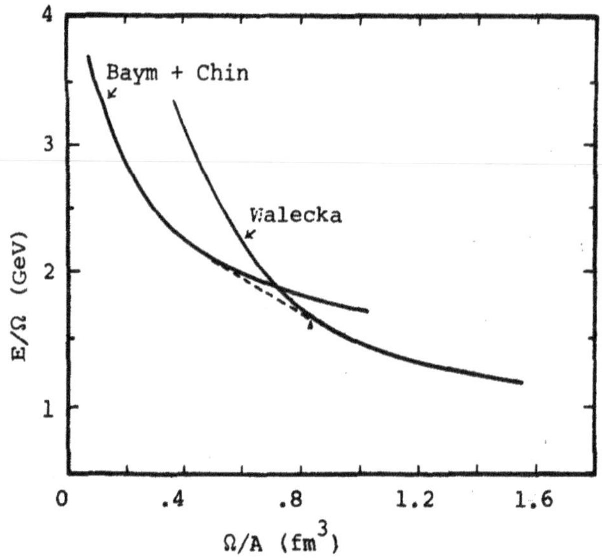

Figrue 18-1. Plots of the equations of state for neutron matter (Walecka) and quark matter (Baym + Chin). The tangent construction method is used to obtain the transition density.

and Walecka (1974). Only one of these is shown in Figure 18-1. They represent two extreme estimates of the neutron matter with most of the others falling between them. The transition densities from neutron matter to quark matter are found to be (i) 1.6×10^{15} g/cm^3 based on Walecka's results, and (ii) 4.1×10^{15} g/cm^3 based on Pandharipande's results. The parameters B and α_c are taken from the MIT bag model with $B \simeq 44$ MeV/fm^3 and $\alpha_c \simeq\simeq 2.2$.

QCD being a renormalizable theory possessing the property of asymptotic freedom, its renormalized coupling constant decreases with increasing energy, density and temperature. It therefore has a meaningful perturbative expansion when the coupling constant is sufficiently small. The perturbative theory of QCD has been well-developed. Since we have not followed the perturbative approach here in this book, we have not developed the methodology in dealing with renormalization or computing terms of higher orders in the coupling constant. The interaction energy computed in (18.21) has the diagrammatic interpretation (of perturbation theory) shown in Figure 18-2(a). It is in fact the perturbative result to order g^2. It may be written in a more suggestive form as follows:

$$E_I/\Omega = \frac{1}{2}(2\pi)^{-6}\int\frac{d^3p}{2p_0}\frac{d^3q}{2q_0}\sum_{\substack{a,b\\ \alpha,\alpha'}}\sum_{\text{spin}}\frac{i\gamma_\nu p_\nu+m}{p^2+m^2}\gamma_\mu\frac{g}{2}\lambda^\alpha_{ab}\frac{i\gamma_\nu q_\nu+m}{q^2+m^2}\gamma_\mu\frac{g}{2}\lambda^{\alpha'}_{ba}\frac{1}{m_a^2+(p-q)^2} \tag{18.23}$$

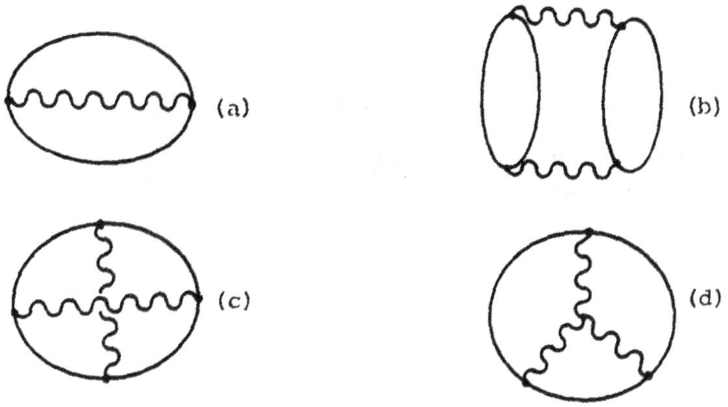

Figure 18-2. Perturbative diagrams contributing to quark interaction energy.

since $u(p)u^{+}(p)$ may be converted into propagators:

$$u(\vec{p})u^{+}(\vec{p}) = \frac{1}{2p_0}\,\frac{i\gamma_\nu p_\nu + m}{p^2+m^2}\,\gamma_4\ . \tag{18.24}$$

To order g^4, the diagrams (b), (c) and (d) shown in Figure 18-2 are to be included. The resulting energy density is found to be:

$$E/\Omega = \frac{3}{4\pi^2}\,p_F^4\,N_f\left(1 + \frac{2\alpha_c}{3\pi} + \frac{\alpha_c^2}{3\pi^2}\{N_f\ \ell n\,\frac{\alpha_c N_f}{\pi} + 0.02\,N_f + 6.75\}\right), \tag{18.25}$$

where N_f denotes the number of flavor of quarks composing the system. The equation of state of quark matter computed to order g^4 is listed in Appendix C(8).

References

Baym, G. and Chin, S.A. (1976). Phys. Lett. 62B, 241.

Lee, T.D. (1981). Particle Physics and Introduction to Field Theory, Harwood Academic, New York.

Pandharipande, V.R. (1971). Nucl. Phys. A178, 123.

Walecka, J.D. (1974). Ann. Phys. 83, 491.

Bibliography

Baluni, V. (1978). Phys. Rev. D17, 2092.

Baym, G. (1978). In Nuclear Physics with Heavy Ions and Mesons, (Balian, R., Rho, M. and Ripka, G., Eds.), Vol. 2, p.745, North-Holland, Amsterdam and New York.

Chodos, A., Jaffe, R,L., Johnson, K., Thorn, C,B. and Weisskopf, V.F. (1974). Phys. Rev. D9, 3471.

Close, F.C. (1979). An Introduction to Quarks and Partons, Academic Press, New York.

Freedman, B.A. and McLerran, L.D. (1977). Phys. Rev. D16, 1130, 1147, 1169.

Kislinger, M.B. and Morley, P.D. (1979). Phys. Rep. 51, 64.

Kokkedee, J.J.J. (1969). The Quark Model, Benjamin, New York.

Rho, M. (1981). In Progess in Particle and Nuclear Physics, (D. Wilkinson, Ed.), Vol. 6, p. 87, Pergamon Press, Oxford.

Shuryak, E.V. (1978). Ah. E.T.F. 74, 408; Sov. Phys. JETP 47, 212.

Shuryak, E.V. (1980). Phys. Rep. 61, 71.

Appendix A

Physical Constants and Astronomical Parameters*

Speed of light	$c = 2.997925 \times 10^{10}$ cm/sec
Gravitational constant	$G = 6.6732 \times 10^{-8}$ dy-cm^2/g^2
	$G/c^2 = 7.425 \times 10^{-29}$ cm/g
Planck's constant	$\hbar = 6.582183 \times 10^{-16}$ eV-sec
	$= 1.054592 \times 10^{-27}$ erg-sec
	$h = 2\pi\hbar = 6.625 \times 10^{-27}$ erg-sec
	$\hbar c = 1.973289 \times 10^{-11}$ MeV-cm
Electron volt	$1\ \text{eV} = 1.6021917 \times 10^{-12}$ erg
Electronic charge (unrat.)	$e = 4.803250 \times 10^{-10}$ esu
Fine structure constant, $\alpha = e^2/\hbar c$	$= 1/137.03602$
Electron mass	$m_e = 9.109558 \times 10^{-28}$ g
	$m_e c^2 = 0.5110041$ MeV
Proton mass	$m_p = 1.67 \times 10^{-24}$ g
	$m_p c^2 = 938.2592$ MeV
Neutron mass	$m_n c^2 = 939.5527$ MeV
Boltzmann constant	$k_B = 1.380622 \times 10^{-16}$ erg/$^\circ$K
	$k_B^{-1} = 11604.85$ $^\circ$K/eV
Solar mass	$M_\odot = 1.989 \times 10^{33}$ g
	$M_\odot G/c^2 = 1.475$ km
Solar radius	$R_\odot = 6.9598 \times 10^5$ km

*Taken from S. Weinberg, Gravitation and Cosmology, John Wiley & Sons, New York, 1972.

Appendix B

Coordinates and Special Functions

B.1 Spherical Polar Coordinates

The Cartesian coordinates fo the point r are denoted by (x,y,z), and its spherical polar coordinates by (r,θ,ϕ), where the colatitude θ and the azimuthal angle ϕ are shown in Figure B-1. They are given by:

$$z = r\cos\theta ,$$
$$x = r\sin\theta\cos\phi ,$$
$$y = r\sin\theta\sin\phi . \qquad (B.1.1)$$

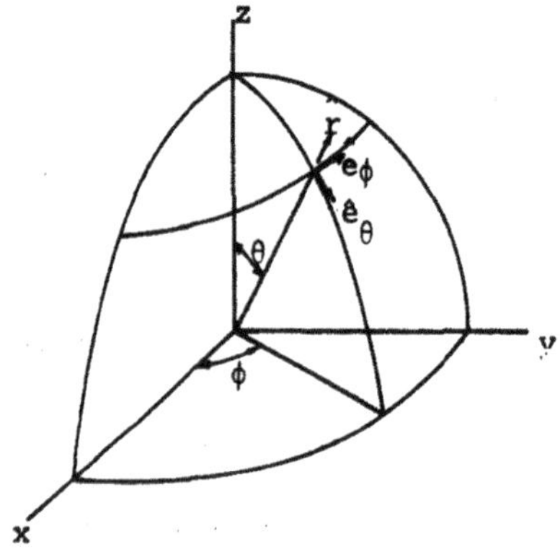

Figure B-1. Cartesian and Spherical Polar Coordinates.

The unit vectors along the x, y, and z directions are indicated by $\hat{e}_x$, $\hat{e}_y$, and $\hat{e}_z$, respectively, and those along the r, θ, and ϕ directions by $\hat{r}$, $\hat{e}_\theta$, and $\hat{e}_\phi$, respectively. They both define orthogonal coordinate systems. In particular,

$$\hat{r} \times \hat{e}_\theta = \hat{e}_\phi , \quad \hat{e}_\theta \times \hat{e}_\phi = \hat{r} , \quad \hat{e}_\phi \times \hat{r} = \hat{e}_\theta . \qquad (B.1.2)$$

In the spherical polar coordinates the differential operators take on the following forms:

gradient: $$\vec{\nabla} = \hat{r}\frac{\partial}{\partial r} + \hat{e}_\theta \frac{1}{r}\frac{\partial}{\partial\theta} + \hat{e}_\phi \frac{1}{r\sin\theta}\frac{\partial}{\partial\phi} , \qquad (B.1.3)$$

divergence: $$\vec{\nabla}\cdot\vec{F} = r^{-2}\frac{\partial}{\partial r}(r^2F_r) + (r\sin\theta)^{-1}\frac{\partial}{\partial\theta}(\sin\theta\, F_\theta) + (r\sin\theta)^{-1}\frac{\partial}{\partial\phi}F_\phi \qquad (B.1.4)$$

curl: $$\vec{\nabla}\times\vec{F} = \frac{1}{r^2\sin\theta}\begin{vmatrix} \hat{r} & r\hat{e}_\theta & r\sin\theta\,\hat{e}_\phi \\ \frac{\partial}{\partial r} & \frac{\partial}{\partial\theta} & \frac{\partial}{\partial\phi} \\ F_r & rF_\theta & r\sin\theta F_\phi \end{vmatrix} , \qquad (B.1.5)$$

(The vertical bars denote a determinant.)

Laplacian: $$\nabla^2 = r^{-2}\frac{\partial}{\partial r}(r^2\frac{\partial}{\partial r}) - r^{-2}\frac{1}{\sin\theta}\frac{\partial}{\partial\theta}(\sin\theta\frac{\partial}{\partial\theta}) - (r\sin\theta)^{-2}\frac{\partial^2}{\partial\phi^2} , \qquad (B.1.6)$$

angular momentum operators:

$$\vec{\ell} = -i\hbar\,(\vec{r}\times\vec{\nabla}) = -i\hbar\left(\hat{e}_\phi\frac{\partial}{\partial\theta} - \hat{e}_\theta\frac{1}{\sin\theta}\frac{\partial}{\partial\phi}\right), \qquad (B.1.7)$$

$$\ell_\pm = \ell_x \pm \ell_y = \pm\hbar\, e^{\pm i\phi}\left(\frac{\partial}{\partial\theta} \pm i\cot\theta\,\frac{\partial}{\partial\phi}\right), \qquad (B.1.8)$$

$$\ell_z = -i\hbar\frac{\partial}{\partial\phi}, \qquad (B.1.9)$$

$$\ell^2 = \ell_z^{\,2} + \frac{1}{2}(\ell_+\ell_- + \ell_-\ell_+)$$

$$= -\hbar^2\left\{(\sin\theta)^{-1}\frac{\partial}{\partial\theta}\left(\sin\theta\frac{\partial}{\partial\theta}\right) + (\sin\theta)^{-2}\frac{\partial^2}{\partial\phi^2}\right\}, \qquad (B.1.10)$$

or,

$$\nabla^2 = r^{-2}\frac{\partial}{\partial r}r^2\frac{\partial}{\partial r} - \frac{\ell^2}{\hbar^2 r^2}. \qquad (B.1.11)$$

The Pauli matrices:

$$\vec{\sigma} = \hat{r}\sigma_r + \hat{e}_\theta\sigma_\theta + \hat{e}_\phi\sigma_\phi \qquad (B.1.11)$$

where

$$\sigma_r = \begin{pmatrix}\cos\theta & \sin\theta e^{-i\phi}\\ \sin\theta e^{i\phi} & -\cos\theta\end{pmatrix} \quad \sigma_\theta = \begin{pmatrix}-\sin\theta & \cos\theta e^{-i\phi}\\ \cos\theta e^{i\phi} & \sin\theta\end{pmatrix} \quad \sigma_\phi = \begin{pmatrix}0 & -ie^{-i\phi}\\ ie^{i\phi} & 0\end{pmatrix}$$

Their commutation relations are:

$$[\,\ell_\theta, \sigma_r\,] = i\hbar\,\sigma_\phi, \qquad [\,\ell_\phi, \sigma_r\,] = -i\hbar\sigma_\theta,$$

$$[\,\sigma_\theta, \sigma_r\,] = -2i\sigma_\phi, \qquad [\,\sigma_\phi, \sigma_r\,] = 2i\sigma_\theta. \qquad (B.1.12)$$

B.2. Spherical Harmonics

The spherical harmonics form a complete basis functions for the angular momentum operators. They satisfy:

$$\ell^2 \, Y_{\ell,m}(\hat{r}) = \hbar^2 \, \ell(\ell+1) \, Y_{\ell,m}(\hat{r}) \,, \tag{B.2.1}$$

$$\ell_z \, Y_{\ell,m}(\hat{r}) = \hbar m \, Y_{\ell,m}(\hat{r}) \,, \tag{B.2.2}$$

where ℓ are non-negative integers and m are integers bounded by:

$$m^2 \leq \ell(\ell+1) \,. \tag{B.2.3}$$

The unit vector $\hat{r}$ in the argument denotes the pair of spherical polar coordinates (θ,ϕ) with which $\hat{r}$ is oriented. Furthermore,

$$\ell_\pm \, Y_{\ell,m}(\hat{r}) = \hbar \sqrt{\ell(\ell+1) \mp m(m\pm1)} \; Y_{\ell,m\pm1}(\hat{r}) \,. \tag{B.2.4}$$

They are orthogonal and complete functions in the sense that:

$$\int_0^{2\pi} d\phi \int_0^{\pi} \sin\theta d\theta \; Y^*_{\ell,m}(\theta,\phi) \; Y_{\ell',m'}(\theta,\phi) \;=\; \delta(\ell,\ell')\delta(m,m') \,, \tag{B.2.5}$$

$$\begin{aligned} \sum_{\ell=0}^{\infty} \sum_{m=-\ell}^{\ell} Y^*_{\ell,m}(\theta',\phi') \; Y_{\ell,m}(\theta,\phi) \;&=\; \delta(\phi-\phi')\delta(\cos\theta-\cos\theta') \\ &=\; \delta(\phi-\phi')\delta(\theta-\theta')/\sin\theta \,, \end{aligned} \tag{B.2.6}$$

where,

$$Y^*_{\ell,m}(\theta,\phi) \;=\; (-1)^m \; Y_{\ell,-m}(\theta,\phi) \,, \tag{B.2.7}$$

and

$$Y_{\ell,-m}(\theta,\phi) \;=\; (-1)^{\ell+m} \; Y_{\ell,m}(\pi-\theta,\pi+\phi) \,. \tag{B.2.8}$$

Explicitly,

$$Y_{\ell,m}(\theta,\phi) = (-1)^m \left(\frac{(2\ell+1)\,(\ell-|m|)!}{4\pi \;\; (\ell+|m|)!} \right)^{1/2} P_\ell^{|m|}(\cos\theta) \; e^{im\phi} \,, \tag{B.2.9}$$

where

$$P_\ell^m(z) \;=\; (1-z^2)^{m/2} \, \frac{d^m}{dz^m} P_\ell(z) \,, \tag{B.2.10}$$

with

$$P_\ell(z) \;=\; \frac{1}{2^\ell \; \ell!} \, \frac{d^\ell}{dz^\ell} (z^2-1)^\ell \,. \tag{B.2.11}$$

The $P_\ell(z)$ and $P_\ell^m(z)$ are called the Legendre polynomials and the associated Legendre polynomials, respectively. They are orthogonal functions:

$$\int_{-1}^{1} dz\, P_\ell^m(z)\, P_{\ell'}^m(z) = \frac{2}{2\ell+1} \frac{(\ell+m)!}{(\ell-m)!} \delta(\ell,\ell') \quad , \tag{B.2.12}$$

and

$$\int_{-1}^{1} dz(1-z^2)^{-1} P_\ell^{m'}(z)\, P_\ell^m(z) = \frac{1}{m} \frac{(\ell+m)!}{(\ell-m)!} \delta(m,m') \tag{B.2.13}$$

A very important relationship among these functions is the addition theorem, which in the present notations is written as:

$$P_\ell(\hat{r}_1,\hat{r}_2) = \frac{4\pi}{2\ell+1} \sum_{m=-\ell}^{\ell} Y_{\ell,m}(\hat{r}_1)\, Y^*_{\ell,m}(\hat{r}_2) \quad . \tag{B.2.14}$$

This is a reduced form of the group properties satisfied by the rotation functions composing of the spherical harmonics. See Edmonds* (1957) for a proof of (B.2.14). Some of the low ℓ spherical harmonics are:

$$Y_{0,0} = (1/4\pi)^{\frac{1}{2}} \quad ,$$

$$Y_{1,0} = (3/4\pi)^{\frac{1}{2}} \cos\theta \quad ,$$

$$Y_{1,\pm1} = \mp(3/8\pi)^{\frac{1}{2}} \sin\theta\, e^{\pm i\phi} \quad ,$$

$$Y_{2,0} = (5/16\pi)^{\frac{1}{2}} (3\cos^2\theta - 1) \quad ,$$

$$Y_{2,\pm1} = \mp(15/8\pi)^{\frac{1}{2}} \sin\theta \cos\theta\, e^{\pm i\phi} \quad ,$$

$$Y_{2,\pm2} = (15/32\pi)^{\frac{1}{2}} \sin^2\theta\, e^{\pm 2i\phi} \quad . \tag{B.2.15}$$

* A.R. Edmonds, Angular Momentum in Quantum Mechanics, Princeton Univ. Press, 1957.

B.3. Bessel Functions and Spherical Bessel Functions

Bessel functions are solutions to the Bessel equation:

$$\{ x^2 \frac{d^2}{dx^2} + x \frac{d}{dx} + (x^2 - n^2)\} y = 0 . \qquad (B.3.1)$$

One such solution regular at x=0 is designated by $J_n(x)$. For n an integer,

$$J_{-n}(x) = (-1)^n J_n(x) . \qquad (B.3.2)$$

When n is not an integer, $J_{-n}(x)$ is an independent solution from $J_n(x)$. One usually defines a second independent solution $Y_n(x)$ (not to be confused with the spherical harmonics):

$$Y_n(x) = \frac{\cos n\pi \; J_n(x) - J_{-n}(x)}{\sin(n\pi)} , \qquad (B.3.3)$$

which has a zero Wronskian with J_n. Other solutions with different asymptotic forms are formed by:

$$H_n^{(1)}(x) = J_n(x) + iY_n(x) ,$$

$$H_n^{(2)}(x) = J_n(x) - iY_n(x) . \qquad (B.3.4)$$

They are called the Hankel functions. As $x \to 0, J_n(x)$ with $n \geq 0$ is finite. All others are infinite at the origin. Asymptotic forms of the solutions as $x \to \infty$ are:

$$J_n(x) \to (2/\pi x)^{\frac{1}{2}} \cos(x - \frac{1}{2}n\pi - \frac{1}{4}\pi)$$

$$Y_n(x) \to (2/\pi x)^{\frac{1}{2}} \sin(x - \frac{1}{2}n\pi - \frac{1}{4}\pi)$$

$$H_n^{(1)}(x) \to (2/\pi x)^{\frac{1}{2}} \exp\{i(x- \frac{1}{2}n\pi - \frac{1}{4}\pi)\}$$

$$H_n^{(2)}(x) \to (2/\pi x)^{\frac{1}{2}} \exp\{-i(x - \frac{1}{2}n\pi - \frac{1}{4}\pi)\}. \qquad (B.3.5)$$

For n given by half-integers, it is convenient to introduce the spherical Bessel functions by:

$$j_n(x) = (\pi/2x)^{\frac{1}{2}} J_{n+\frac{1}{2}}(x) ,$$

$$n_n(x) = (\pi/2x)^{\frac{1}{2}} Y_{n+\frac{1}{2}}(x) , \qquad (B.3.6)$$

and the spherical Hankel functions by:

$$h_n^{(1,2)}(x) = j_n(x) \pm i n_n(x) . \qquad (B.3.7)$$

We shall use the notations: $h_n = h_n^{(1)}$, and $h_n^* = h_n^{(2)}$ here. They are solutions to the equation:

$$\{\frac{d^2}{dr^2} + \frac{1}{r}\frac{d}{dr} + \left(k^2 - \frac{(n+\frac{1}{2})^2}{r^2}\right)\}y = 0 , \qquad (B.3.8)$$

with $x = kr$. The spherical Bessel functions are expressible in terms of the trigonometric functions:

$$j_n(x) = (-x)^n \left(\frac{1}{x}\frac{d}{dx}\right)^n \{\frac{\sin(x)}{x}\},$$

$$n_n(x) = (-x)^n \left(\frac{1}{x}\frac{d}{dx}\right)^n \{-\frac{\cos(x)}{x}\} . \qquad (B.3.9)$$

Asymptotically, as $x \to \infty$:

$$j_n(x) \to \frac{1}{x}\cos\left(x - \frac{1}{2}\pi(n+1)\right) ,$$

$$n_n(x) \to \frac{1}{x}\sin\left(x - \frac{1}{2}\pi(n+1)\right) ,$$

$$h_n(x) \to \frac{1}{x}\exp\{i\left(x - \frac{1}{2}\pi(n+1)\right)\}. \qquad (B.3.10)$$

Explicit expressions of the low n spherical Bessel functions are:

$$j_0(x) = \frac{\sin x}{x}, \quad j_1(x) = \frac{\sin x}{x^2} - \frac{\cos x}{x}, \quad j_2(x) = \left(\frac{3}{x^3} - \frac{1}{x}\right)\sin x - \frac{3\cos x}{x^2},$$

$$n_0(x) = -\frac{\cos x}{x}, \quad n_1(x) = -\frac{\cos x}{x^2} - \frac{\sin x}{x}, \quad n_2(x) = -\left(\frac{3}{x^3} - \frac{1}{x}\right)\cos x - \frac{3\sin x}{x^2},$$

$$h_0(x) = \frac{e^{ix}}{ix}, \quad h_1(x) = -\left(1 - \frac{i}{x}\right)\frac{e^{ix}}{x}, \quad h_2(x) = \left(1 + \frac{3i}{x} - \frac{3}{x^2}\right)\frac{i\,e^{ix}}{x} .$$

(B.3.11)

Appendix C

Equation of State of Dense Matter

Samples of the equation of state of dense matter in its ground state and their numerical values are compiled here. Some are better determined than the others (see text). In some density ranges, several versions are presented for comparison. A similar listing of the equation of state based on studies carried out before 1974 is given by Canuto (1974, 1975).

A. Equation of state at T = 0.

(1) Baym, Pethick, Sutherland (1971)

ρ (g/cm^3)	P (dy/cm^2)
2.120E2	5.82 E15
1.150E3	1.90 E17
1.044E4	9.744E18
2.622E4	4.968E19
6.587E4	2.431E20
1.654E5	1.151E21
4.156E5	5.266E21
1.044E6	2.318E22
2.622E6	9.755E22
6.588E6	3.911E23
8.293E6	5.259E23
1.655E7	1.435E24
3.302E7	3.833E24
6.589E7	1.006E25
1.315E8	2.604E25
2.624E8	6.676E25
3.304E8	8.738E25
5.237E8	1.629E26
8.301E8	3.029E26
1.045E9	4.129E26
1.316E9	5.036E26
1.657E9	6.860E26
2.626E9	1.272E27
4.164E9	2.356E27
6.601E9	4.362E27
8.312E9	5.662E27
1.046E10	7.702E27
1.318E10	1.048E28
1.659E10	1.425E28
2.090E10	1.938E28
2.631E10	2.503E28
3.313E10	3.404E28
4.172E10	4.628E28

ρ (g/cm^3)	P (dy/cm^2)
5.245E10	5.949E28
6.617E10	8.089E28
8.332E10	1.100E29
1.049E11	1.495E29
1.322E11	2.033E29
1.664E11	2.597E29
1.844E11	2.892E29
2.096E11	3.290E29
2.640E11	4.473E29
3.325E11	5.816E29
4.188E11	7.538E29
4.299E11	7.805E29

(2) Baym, Bethe, Pethick (1971)

4.460E11	7.890E29
5.228E11	8.352E29
6.610E11	9.098E29
7.964E11	9.831E29
9.728E11	1.083E30
1.196E12	1.218E30
1.471E12	1.399E30
1.805E12	1.638E30
2.202E12	1.950E30
2.930E12	2.592E30
3.833E12	3.506E30
4.933E12	4.771E30
6.248E12	6.481E30
7.801E12	8.748E30
9.611E12	1.170E31
1.246E13	1.695E31
1.496E13	2.209E31
1.778E13	2.848E31
2.210E13	3.931E31
2.988E13	6.178E31
3.767E13	8.774E31
5.081E13	1.386E32
6.193E13	1.882E32
7.732E13	2.662E32
9.826E13	3.897E32
1.262E14	5.861E32
1.586E14	8.595E32
2.004E14	1.286E33
2.520E14	1.900E33
2.761E14	2.242E33
3.085E14	2.751E33

(3) Neutron matter: Friedman, Pandharipande (1981)

ρ (g/cm^3)	P (dy/cm^2)
1.22E13	1.70E31
1.93E13	3.38E31
2.89E13	6.19E31
4.11E13	1.01E32
5.64E13	1.52E32
7.57E13	2.24E32
9.75E13	3.36E32
1.24E14	5.26E32
1.55E14	8.60E32
1.90E14	1.44E33
2.31E14	2.30E33
2.77E14	3.85E33
3.21E14	5.71E33
3.71E14	8.37E33
4.30E14	1.24E34
4.97E14	1.82E34
5.76E14	2.72E34
6.55E14	4.06E34
7.72E14	6.22E34
8.94E14	9.63E34
1.03E15	1.30E35
1.23E15	2.21E35

(4) Neutron matter: Walecka (1974)

1.15E15	2.83E35
1.60E15	5.47E35
2.13E15	8.76E35
3.06E15	1.48E36
4.55E15	2.50E36
6.32E15	3.76E36
8.36E15	5.24E36
1.07E16	6.95E36
1.33E16	8.89E36
1.61E16	1.11E37
1.93E16	1.35E37
2.26E16	1.61E37
2.63E16	1.89E37
3.02E16	2.20E37
3.44E16	2.53E37

(5) Baryonic matter: Canuto 1974)

2.20E14	1.51E33
2.55E14	2.15E33
2.89E14	2.96E33
3.23E14	3.98E33
3.58E14	5.22E33

ρ (g/cm^3)	P (dy/cm^2)
3.93E14	6.71E33
4.27E14	8.47E33
4.80E14	1.17E34
5.51E14	1.71E34
6.23E14	2.40E34
6.95E14	3.25E34
7.87E14	4.57E34
8.82E14	6.20E34
9.78E14	8.17E34
1.08E15	1.05E35
1.28E15	1.64E35
1.49E15	2.41E35
1.72E15	3.38E35
1.96E15	4.57E35
2.22E15	5.99E35
2.63E15	8.62E35
3.08E15	1.19E36

(6) Baryonic matter: Malone, Johnson, Bethe (1975)

1.67E14	1.19E33
2.51E14	2.93E33
3.34E14	6.00E33
4.18E14	1.09E34
5.01E14	1.83E34
6.68E14	4.09E34
8.35E14	7.61E34
1.00E15	1.26E35
1.17E15	1.99E35
1.34E15	2.85E35
1.54E15	3.71E35
1.67E15	4.02E35
1.84E15	5.02E35
2.09E15	6.76E35
2.34E15	8.81E35
2.51E15	1.03E36
2.84E15	1.38E36
3.34E15	2.02E36
4.18E15	3.40E36
5.01E15	5.20E36

(7) Pion condensed matter: Weise, Brown (1975)

8.35E13	4.56E32
1.84E14	1.66E33
2.51E14	3.01E33
3.51E14	6.24E33
4.18E14	9.47E33
4.61E14	1.19E34
6.68E14	1.19E34
7.52E14	1.55E34
9.19E14	2.94E34

ρ (g/cm³)	P (dy/cm²)
1.00E15	6.53E34
1.34E15	9.79E34
1.50E15	1.35E35
1.67E15	1.78E35
1.84E15	2.27E35

(8) Quark Matter; Morley, Kislinger (1979)

4.33E14	6.73E34
4.66E14	7.45E34
5.02E14	8.22E34
5.39E14	9.05E34
5.79E14	9.92E34
6.21E14	1.09E35
6.65E14	1.18E35
8.33E14	1.26E35
9.05E14	1.35E35
9.64E14	1.43E35
1.02E15	1.51E35
1.08E15	1.60E35
1.14E15	1.69E35
1.31E15	1.98E35
1.70E15	2.66E35
2.50E15	4.16E35
3.01E15	5.20E35
3.99E15	7.27E35
4.96E15	9.39E35
5.99E15	1.17E36
6.99E15	1.40E36
7.81E15	1.60E36
8.69E15	1.81E36
9.54E15	2.01E36
1.01E16	2.14E36

B. Equation of state along an adiabat.

(9) Subnuclear matter, Y_p=0.3, (s/k_B)= 1.5: Brown, Bethe, Baym (1982)

ρ (g/cm³)	P (dy/cm²)	k_BT (MeV)
2.65E11	4.65E29	1.31
2.67E12	1.04E31	2.59
2.71E13	2.23E32	5.48
4.31E13	4.07E32	6.37
6.87E13	7.42E32	7.38
10.97E13	1.36E33	8.50
17.44E13	2.90E33	9.20

(10) Subnuclear matter, Y_p=0.25, (s/k_B)=1: Bonche, Vautherin (1981)

ρ (g/cm^3)	P (dy/cm^2)	$k_B T$ (MeV)
3.34E13	2.17E32	3.80
5.01E13	3.69E32	4.35
6.68E13	5.37E32	4.15
8.35E13	7.17E32	5.06
1.00E14	9.05E32	5.34
1.17E14	1.10E33	5.53

(11) Transnuclear matter, Y_p=0.3, (s/k_B)=1.1: Brown, Bethe, Baym (1982)

3.0E14	6.56E33	7.5
4.5E14	1.63E34	9.8
6.0E14	3.36E34	11.9
7.5E14	3.48E34	13.8
9.0E14	10.88E34	15.6

References

Baym, G., Bethe, H.A. and Pethick, C.J. (1971). Nucl. Phys. A175, 225.

Baym, G., Pethick, C.J. and Sutherland, P. (1971). Astrophys. J. 170, 299.

Bonche, P. and Vautherin, D. (1981). Nucl. Phys. A372, 496.

Brown, G.E., Bethe, H.A. and Baym, G. (1982). Nucl. Phys. A375, 481.

Canuto, V. (1974). Ann. Rev. Astron. Astrophys. 12, 167.

Canuto, V. (1975). Ann. Rev. Astron. Astrophys. 13, 355.

Friedman, B. and Pandharipande, V.R. (1981). Nucl. Phys. A361, 502.

Malone, R.C., Johnson, M.B. and Bethe, H.A. (1975). Astrophys. J. 199, 741.

Walecka, J.D. (1974). Ann. Phys. 83, 491.

Exercises

(Exercises are listed according to sections; number in front of the dashed line is the section number)

Ex 1-1. Write out the Hartree-Fock trial wave function for $N = 3$.

Ex 1-2. Derive the factors b_i and d_{ij} of (1.21) and (1.22) for $N = 3$.

Ex 1-3. Write out the Hartree-Fock equations (1.25) for $N = 3$

Ex 2-1. Apply the method of separation of variables to (2.1) to obtain the corresponding one-dimensional equations. Relate ε_i to the eigenvalues of the one-dimensional equations.

Ex 2-2. Derive ε_K and P of (2.21) and (2.22).

Ex 2-3. Show that for an ideal degenerate electron gas the equation of of state approaches $P = \frac{1}{3}\varepsilon_K$ in the extreme relativistic limit $e_K \to pc$. Estimate the electron number density for which the above equation of state is accurate to 1%. What would be the corresponding density of a helium gas?

Ex 2-4. Obtain the numerical values of pressure for a noninteracting helium plasma at densities of 10^4, 10^5, 10^6 and 10^7 g/cm^3.

Ex 2-5. Obtain the numerical values of the adiabatic index for Ex 2-4.

Ex 3-1. A solution to the Thomas-Fermi equation (3.13) for metallic Cu (Z=29) is found to be $u_s = 0.0599$ at $x_s = 9.565$.

(a) Evaluate the electron number density n_{TF} of (3.16) at that point.

(b) Evaluate the pressure (3.17) exerted by the electron gas at x_s.

(c) Evaluate the density of the metal.

Ex 3-2. Derive the expansion coefficients (a_2 to a_9) of the Thomas-Fermi-Dirac function w of (3.29).

Ex 3-3. A solution to the Thomas-Fermi-Dirac equation (3.28) for Uranium is found to be $u_s = 0.01991$ at $x_s = 12.4037$. Evaluate the pressure (3.30) exerted by the electron gas at x_s.

Ex 4-1. A simple harmonic oscillator has energy levels given by $E_n = (n + \frac{1}{2})\hbar\omega$, where the quantum number n = 0, 1, 2, ... Suppose that such an oscillator is in thermal contact with a heat reservoir at temperature T low enough so that $k_B T/(\hbar\omega) << 1$.

(a) Find the ratio of the probability of the oscillator being in the first excited state n = 1 to the probability of its being in the ground state n = 0.

(b) Assuming that only the ground state and first excited state are appreciably occupied, find the mean energy of the oscillator as a function of the temperature T.

Ex 4-2. Consider a system consisting of two Fermi particles, each of which can be in any one of three quantum states of respective energies 0, ε, and 3ε. The system is in contact with a heat reservoir at temperature T. Write an expression for the partition function.

Ex 4-3. Show that for temperature low compared to the maximum kinetic energy, or $k_B T << (Ze^2/a)(u_s/x_s)$, the electron density (cf. 3.16) may be approximated by:

$$n_{TF} = \frac{8\pi}{3h^3}\left(3m(\varepsilon_F+e\Phi)\right)^{1/2}\left[1 + \frac{\pi^2 k_B^2 T^2}{8(\varepsilon_F+e\Phi)}\right]$$

(Ref: Marshak, R.E. and Bethe, H.A. (1940). Astrophys. J. 91, 239)

Ex 5-1. Obtain the equation of state of an electron-proton plasma as its density increases from 10^7 to 10^9 g/cm^3. Assuming in this case the neutrinos are trapped, so that the neutrinos are in equilibrium with the protons, neutrons and electrons. The condition of chemical equilibrium becomes $\mu_n+\mu_\nu = \mu_p+\mu_e$.

Ex 6-1. In the Fermi gas model of the atomic nucleus, the nucleus is approximated by a degenerate noninteracting Fermi gas of neutrons and protons.

(a) What is the degeneracy factor γ for each level?

(b) If the radius of a nucleus with A nucleons is given by (6.1), what are the Fermi momentum k_F and Fermi energy ε_F?

How do they vary with A?

(c) What is the pressure exerted by this Fermi gas?

(d) If each nucleon is considered to be moving in a constant potential of depth V_0, how large must v_0 be?

Ex 6-2. (a) Compute the average energy per nucleus according to (6.19) at $n_B = 10^{-6}$ fm^{-3} for the following nuclei: Fe^{56}, Ni^{62}, Ni^{64} and Se^{84} with nuclear mass given by Table 6-1.

(b) Compute the mass density and pressure for each case.

Ex 6-3. When M(z,A) is greater than M(Z-2,A-4) + M(2,4), the nucleus (Z,A) is unstable against α-emission. Compute the kinetic energy of the emitted α-particle as a function of A, for the range of A for which the energy is positive on the basis of the nuclear mass formula (6.6) with the parameters given by (6.8). Use the most stable charge for Z.

Ex 7-1. (a) Evaluate $\vec{\sigma}_1 \cdot \vec{\sigma}_2 \chi_1^1$.

(b) Evaluate $\vec{\sigma}_1 \cdot \vec{\sigma}_2 \chi_0$.

Ex 7-2. Evaluate $S_{12} \chi_1^1$.

Ex 7-3. (a) Derive the average energy per particle, an expression similar to (7.39), for a pure neutron system.

(b) Obtain the expression for a single-particle potential for a pure neutron system.

Ex 8-1. Work out all the steps leading to the expression for the average energy per nucleon (8.13).

Ex 8-2. Derive the expression for compression modulus K for a system described by (8.24).

Ex 8-3. Compare the pressure for a pure neutron system using the Skyrme potentials with SI and RBP parameters.

Ex 8-4. The isothermal compressibility in thermodynamics is defined to be:

$$K_{th} = -\frac{1}{\Omega}\left(\frac{\partial\Omega}{\partial P}\right)_T$$

Show that the compression modulus K defined in (8.14) is

related to K_{th} at $T = 0$ by:

$$K = \frac{9}{n_B K_{th}} .$$

Ex 9-1. Consider a single spin-½ particle. Show that the operators:

$$\Lambda_\ell^+ = \frac{\ell + 1 + \vec{\sigma}\cdot\vec{\ell}}{2\ell+1}, \qquad \Lambda_\ell^- = \frac{\ell - \vec{\sigma}\cdot\vec{\ell}}{2\ell+1}$$

are projection operators onto the state $j = \ell\pm\frac{1}{2}$, respectively, in the subspace of orbital angular momentum ℓ.

Ex 9-2. Show that the "spherical" components of the vector $\vec{r}$, namely $x \pm iy$ and z, are proportional to the spherical harmonics $Y_{1m}(\theta,\phi)$ with $m = \pm 1, 0$.

Ex 9-3. Verify that the representation of the δ-function in the spherical coordinates is given by:

$$\delta(\vec{r} - \vec{r}\,') = r^{-2}\,\delta(r - r')\,\delta(\cos\theta - \cos\theta')\,\delta(\phi - \phi') ,$$

where (r,θ,ϕ) and (r',θ',ϕ') are the coordinates of $\vec{r}$ and $\vec{r}\,'$. respectively.

Ex 9-4. For ℓ_θ, ℓ_ϕ, σ_r, σ_θ, σ_ϕ defined in Appendix B.1, verify the commutation relations (B.1.12).

Ex 10-1. Evaluate the equation of state of a uniform system of pure neutrons for the density range $10^{14} - 10^{15}$ g/cm^3 at the following temperatures: $k_BT = 0, 10, 20$ MeV, using the RBP Skyrme parameters.

Ex 10-2. Determine the single-nucleon potential V_q of (10.3) for a uniform system of symmetric nuclear matter (Y_p=0.5) in the density range $10^{14} - 10^{15}$ g/cm^3 using the Skyrme potentials with RBP parameters at the following temperatures: $k_BT = 0$, 10, 20 MeV.

Ex 11-1. Express $S_{12}\chi_1^m$ ($m = \pm 1, 0$) in terms of spherical harmonics Y_{1m} and χ_1^m.

Ex 11-2. Show that

$$(\vec{\sigma}_1\cdot\vec{\nabla})(\vec{\sigma}_2\cdot\vec{\nabla})\frac{e^{-\mu r}}{\mu r} = \frac{1}{3}\,\mu^2\,\Big((\vec{\sigma}_1\cdot\vec{\sigma}_2)+S_{12}\{1+\frac{3}{\mu r}+\frac{3}{(\mu r)^2}\}\Big)\frac{e^{-\mu r}}{\mu r} .$$

Ex 11-3. Let us decompose the scattering amplitude $f(\theta)$, which possesses azimuthal symmetry, into partial waves as follows:

$$f(\theta) = \sum_{\ell=0}^{\infty} (2\ell+1)\, f_\ell(k)\, P_\ell(\cos\theta) \;,$$

where k is the momentum of the particle with respect to a fixed potential. Show that if $f_\ell(k)$ is given by real phaseshifts δ_ℓ:

$$f_\ell(k) = \frac{e^{i\delta_\ell}\sin\delta_\ell}{k} \;,$$

then $f(\theta)$ satisfies the following relation:

$$\sigma = \int d\Omega\, |f(\theta)|^2 = \frac{4\pi}{k}\, \mathrm{Im}\, f(\theta=0) \;,$$

where Im stands for the imaginary part and $f(\theta=0)$ is the forward scattering amplitude. This result is called the optical theorem for elastic scattering.

Ex 12-1. Let us write the BBG equation (12.18) in a short-hand notation as follows:

$$G = v + v\frac{Q}{e}G \;.$$

If the two-nucleon potential v may be splitted into two parts,

$$v = v_s + v_\ell$$

where the short-range part v_s contains the strong repulsive core plus the medium range attraction, while v consists of the weak long-range tail. Define the following G-matrices for the short-range potential v_s:

$$G_s = v_s + v_s \frac{1}{e} G_s \;,$$

and

$$G_s^F = v_s + v_s \frac{1}{e_0} G_s^F \;,$$

where e_0 is the free-energy denominator, without the one-body potential U. Note that the Pauli operator Q is absent.

(a) Show that

$$G_s = G_s^F + G_s^F \left(\frac{1}{e} - \frac{1}{e_0}\right) G_s$$

(b) Show that, to second order in v and G_s:

$$G = v_\ell + G_s + G_s\left(\frac{Q-1}{e}\right)G_s + v_\ell \frac{Q}{e} v_\ell + v_\ell \frac{Q}{e} G_s + G_s \frac{Q}{e} v_\ell$$

and to the same order,

$$G = v_\ell + G_s^F + G_s^F \left(\frac{1}{e} - \frac{1}{e_0}\right) G_s^F + G_s^F\left(\frac{Q-1}{e}\right)G_s^F + v_\ell \frac{Q}{e} v_\ell +$$

$$+ v_\ell \frac{Q}{e} G_s^F + G_s^F \frac{Q}{e} v_\ell .$$

(For an explanation to the significance of this approach see Moszkowski and Scott, Ann. Phys. 11, 65, 1960).

Ex 12-2. The tensor operator is defined to be $S_{12}(\vec{r}) = (3\vec{\sigma}_1\cdot\hat{r}\vec{\sigma}_2\cdot\hat{r} - \vec{\sigma}_1\cdot\vec{\sigma}_2)$

(a) Show that $S_{12}^2 = 6 + 2\vec{\sigma}_1\cdot\vec{\sigma}_2 - 2\,S_{12}$.

(b) Let $\psi_{\vec{k}}(\vec{r}) = \frac{1}{\sqrt{\Omega}} \exp(i\vec{k}\cdot\vec{r})$, show that

$$\int d^3r_1 d^3r_2 \psi^*_{\vec{k}_1+\vec{q}}(\vec{r}_1)\psi^*_{\vec{k}_2-\vec{q}}(\vec{r}_2) V_T(|\vec{r}_1-\vec{r}_2|) S_{12}(\vec{r}_1-\vec{r}_2)\psi_{\vec{k}_1}(\vec{r}_1)\psi_{\vec{k}_2}(\vec{r}_2)$$

$$= -\left(\frac{4\pi}{\Omega}\right) S_{12}(\vec{q})\int_0^\infty j_2(qr)V_T(r)r^2dr$$

(c) Take $V_T = V_0\exp(-\mu r)/r$, and estimate the value of q at which the integral in (b) is maximum.

Ex 13-1. Verify that $P_J^{(2)}$ and $P_J^{(3)}$ of (13.24) satisfy the requirement (13.23).

Ex 13-2. Verify the results given by (13.29).

Ex 13-3. Verify the result given by (13.34).

Ex 13-4. Verify the result given by (13.36).

Ex 14-1. Verify that the matrices (14.5) satisfy the algebra of (14.4).

Ex 14-2. (a) Evaluate m_H from (14.45) at $k_F = 1.42\ \text{fm}^{-1}$ ($\hbar = c = 1$) for $\gamma = 4$ with (g_σ/m_σ) given by (14.50).

(b) Evaluate the compression modulus K at this density.

Ex 14-3. (a) Evaluate m_H at $k_F = 1.42\ fm^{-1}$ for $\gamma = 2$ (neutron matter).

(b) Evaluate E/Ω and P at $k_F = 1.42\ fm^{-1}$ for $\gamma = 2$.

Ex 15-1. Derive $D_a(x,t)$ as given by (15.46).

Ex 15-2. Verify relation (15.55).

Ex 16-1. What is the matter density when Λ^o's begin to appear in equilibrium?

Ex 16-2. What is the matter density when Σ^-'s begin to appear in equilibrium?

Ex 17-1. Verify the expression for E'/Ω of (17.17) using conditions (17.18) to (17.22).

Ex 17-2. Insert the results of (17.26) into (17.25) to determine p_{Fn} and p_{Fp}.

Ex 17-3. Verify the derivation of (17.38).

Ex 17-4. Apply S(p) of (17.42) onto (17.40) to show that S(p) is in fact its inverse.

Ex 18-1. Verify the result (18.21).

INDEX

www.ingramcontent.com/pod-product-compliance
Lightning Source LLC
Chambersburg PA
CBHW081533190326
41458CB00015B/5534
* 9 7 8 9 9 7 1 9 7 8 1 0 5 *